Helion & Company Limited
Unit 8 Amherst Business Centre Budbrooke Road, Warwick CV34 5WE, England
Telephone 01926 499 619 • Fax 0121 711 4075 • Email: info@helion.co.uk • Website: www.helion.co.uk
Twitter: @helionbooks • Visit our blog http://blog.helion.co.uk/
Published by Helion & Company 2019
Designed and typeset by Farr out Publications, Wokingham, Berkshire
Cover designed by Paul Hewitt, Battlefield Design (www.battlefield-design.co.uk)
Printed in the UK by Henry Ling Limited, Dorchester, Dorset
Text © Peter Dennis 2019
Pictures © Peter Dennis 2019

The figure artwork pages in this book are copyright free for your personal hobby use. They may not be reproduced for sale or included in other publications without the consent of the copyright holder.

ISBN 978-1-911628-28-6

British Library Cataloguing-in-Publication Data.
A catalogue record for this book is available from the British Library.

All rights reserved. No part of this publication may be reproduced, stored in a retrieval system, or transmitted, in any form, or by any means, electronic, mechanical, photocopying, recording or otherwise, without the express written consent of Helion & Company Limited, other than for personal hobby use.

For details of other military history titles published by Helion & Company Limited contact the above address, or visit our website: http://www.helion.co.uk.

We always welcome receiving book proposals from prospective authors.

MAKING PAPER ARMIES

COPYING

This series of books is intended to be photocopied. They are copyright free for your personal hobby use. They are A4 size pages, which is an international size, but little used in the USA . They will fit on US legal-size paper. If you wish to print on letter-size paper reduce them to 93%. Although you can copy them with your home equipment, I prefer to use a commercial copier, which I think gives a better quality print. Most commercial copiers will allow you to use your own paper. You will need to source thicker card for the base which you will build into the model. BASES are 40mm wide by 30mm deep for all cavalry and three-rank infantry, 40mm by 20mm for two-rank infantry and 50mm wide by 70mm deep for artillery. The 'ground' texture and colour on these figures is by kind permission of Cigar Box Battles. It is their mat 'Open Grassland'.

Some sheets require you to add regimental colour distinctions. Do this first before you score the sheet. Staedtler Triplus Fineliner pens are good for this, but any fine felt tip should do the job. Sharpies are too powerful in colour and hard to control for this use.

PAPER

Most bulk paper is 80 gsm (20lb US) which is a little flimsy, but will work OK. I prefer 100gsm (28lb US) which is still easy to cut out, but gives a stronger result. It is the glue between the layers that gives the model its stiffness though, besides the quality of the paper. You might imagine that even heavier-gauge paper would be better, but it is tiring to cut and the result has a thick white edge which I dislike.

GLUE

If you are in Europe the best glue in my opinion is UHU all-purpose. UHU multi-purpose is a different glue altogether and should be avoided. PVA, the universal white glue, is important in the making of these boys, and can be used as the main stand-making glue, but it is rather wet and may cockle larger subjects like cavalry. Pritt-stick and other stick glues work well but don't give much slide time when matching up front and back views.

In the USA, Scotch 'create' and 'wrinkle-free' stick glues work well. There is also PVA white glue. Ask at the craft shop or hardware store for advice.

PIKES. Fold the colour block in half, open it out and glue the whole back of the sheet. Enfold a sheet of stiff paper inside and press firmly. When thoroughly dry, cut with a sharp craft knife and a ruler into pikes about 1.3 mm thick then sharpen the points

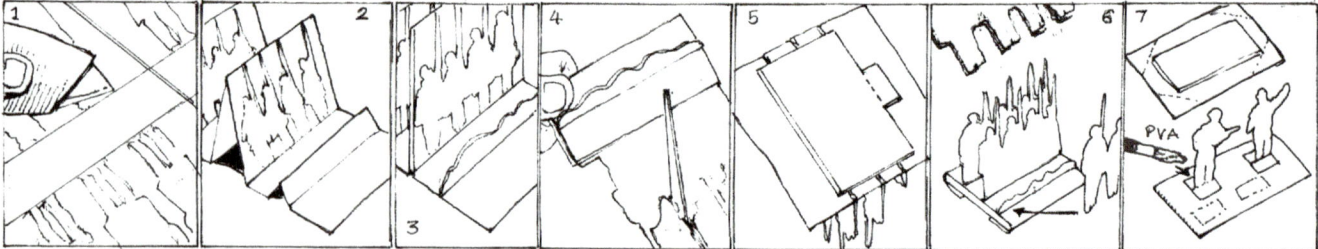

MAKING A STAND

1. Score lightly along the horizontal fold lines of your printed sheet. I do this by hand but you may prefer to use a ruler. Scoring is lightly cutting partway through the paper. You might also use a scoring tool. **2.** Cut the strips out and fold them like the diagram so that the soldiers, and thin 'locator strip' at the front stand up and the ground lies flat. **3.** Glue the upright sheets together. Don't be mean with the glue, it is providing stiffness as well as adhesion. When the assembly is thoroughly dry cut the locator strip into a wavy line. Also glue the separate front and back sheet, which will be the front rank. **4.** Cut out the outline of the ranks attached to the base. Always cut from the front. They will fold conveniently to allow this. Keep your scissor hand steady and move the model you are cutting around it. Cut out the front rank stand and cut out the coloured area between their legs. This makes the stand look a lot better, but there is base colour there so it is optional. **5.** Glue a base to the underside of the stand. Note that in the diagram the base has an extra bit at the front. This is a command stand, and that extra square is for the individual figure to stand on. One stand per battalion should have the command strip with the appropriate flag or flags added as the front rank. **6.** Trim off the extra 'ground' at the sides and fold and glue the front and back 'ground' under the stand to give a neat finish. Glue the front rank to the locator strip. Contact adhesive is best for this. **7.** For skirmish strips, individual commanders and artillery bases, glue your card base to the centre of the back of the ground square, trim the edges and fold and glue. Individuals, like artillery crew, are cut out with their bases 'open' and then glued into place where you wish, square-on with the base front. Add a lick of PVA white glue or mod-podge on a brush to the back of muskets and projecting weapons and also to the back of the heels of the men. This will dry invisibly and add an important element of stiffness to the finished stand. Do this to every stand you make, it only takes a moment and makes a big difference! There is a YouTube video of me making a stand of soldiers. 'Peter Dennis Paper Soldiers' should find it.

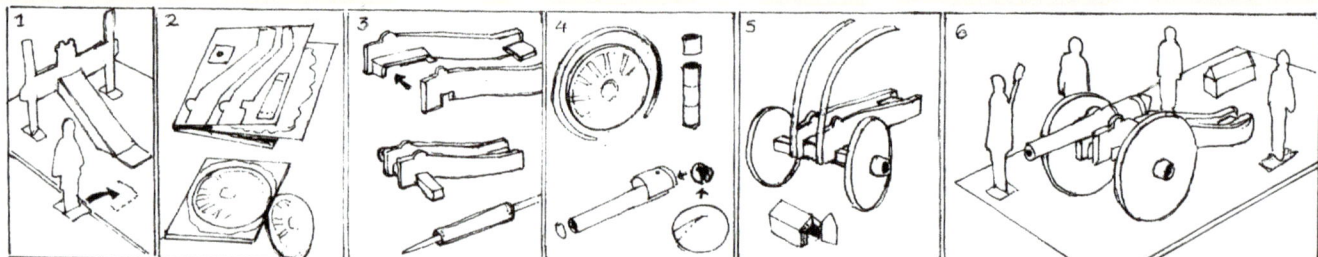

MAKING ARTILLERY

1. If you wish, you can use the 2D artillery system which gives a convincing view of a cannon from the front. The 3D artillery is more solid-looking from all angles. It does take more time and patience to make, but I think it's worth the effort. Before you start, turn over the sheet and blacken the back of the tyre strips and wheel hubs, this will save painting them later. **2.** Fold the trail section in half and glue some thin card on both sides and set it between the trail sheets. Cut out one side of the wheel really carefully and roughly cut the other, leaving them joined. Glue the rough side to similar card, making sure the other wheel side can fold exactly underneath. **3.** While all that is drying make the axle box and roll the cannon barrel around something like a cook's needle or thin knitting needle. Glue the barrel down one edge when you are happy with the tightness. Set aside to dry. Cut out the trail pieces with scissors or a craft knife and assemble it as in the diagram. When dry, slot in the axle. **4.** Carefully cut out the glued wheel half, then glue on the other side of the wheel. Cut the tyre strips and glue round the edge. Blackened inside edges will show but they look quite neat. Roll and glue the hub strip round the needle you use for the barrel. Add the breech piece, making sure the join is in the same place as on the barrel. Make the tiny cone for the rear closure then trim it to size when glued in place and dry. Glue the flat muzzle disc too, when you have cut the barrel to the length you want. **5.** Glue the trail edging strips to the front of the trail and allow to dry. This makes it very easy to glue the rest down the tops of the trail sides. Add the barrel to the trail and the hubs cut to length to the wheels and the gun is finished. 6. Make the base as in pic 7 above and glue the ammo box, and the crew around the gun with their bases square-on to the front.

'The Paperboys Page' on Facebook is the place for any queries you may have and to see the latest work on new Paperboys subjects. Many free downloads are available from the Paper Soldier section of the Helion & Co website.

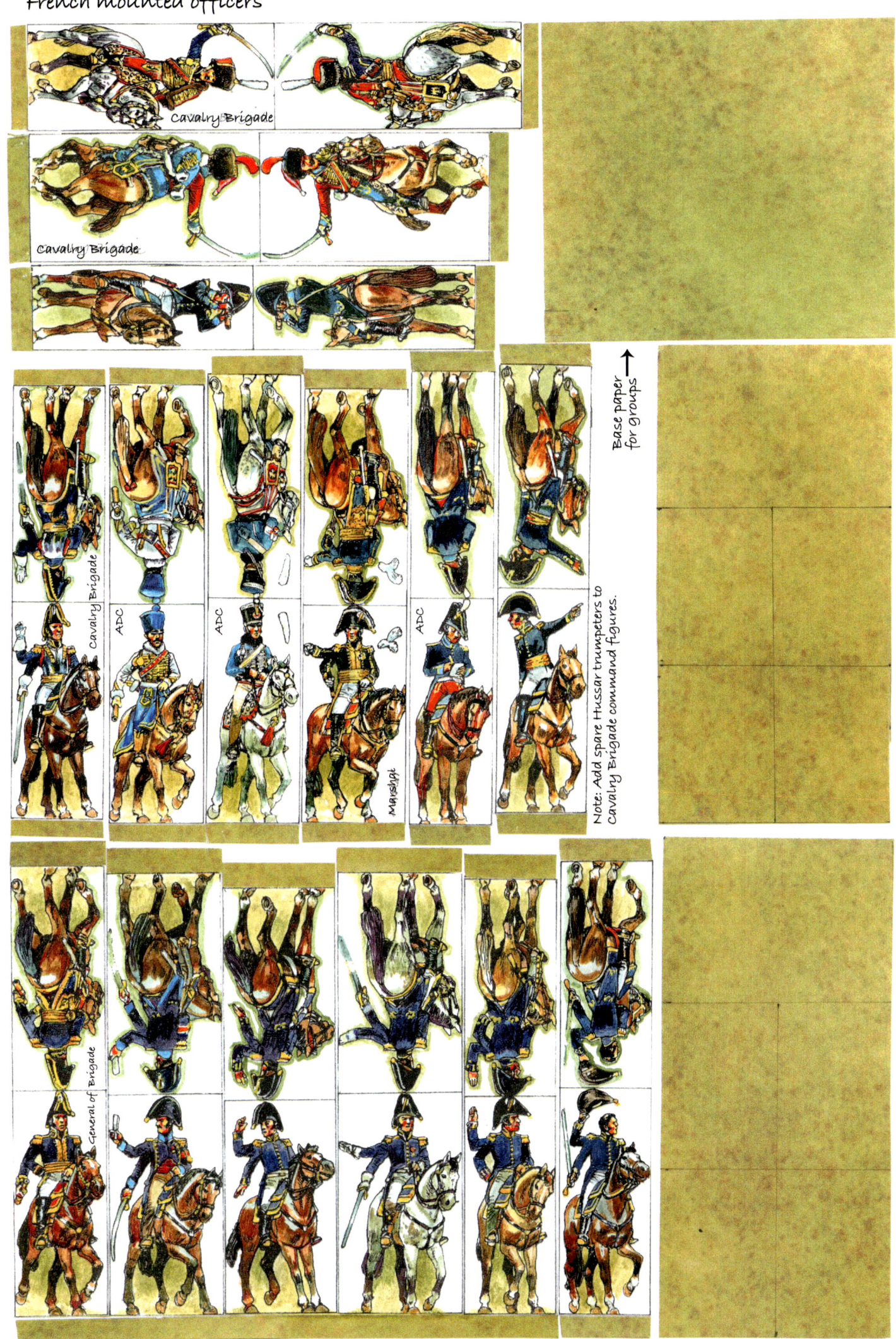

French Line Infantry centre companies

Skirmish bases on flank company sheet

French Line Infantry Voltigeurs and Grenadiers

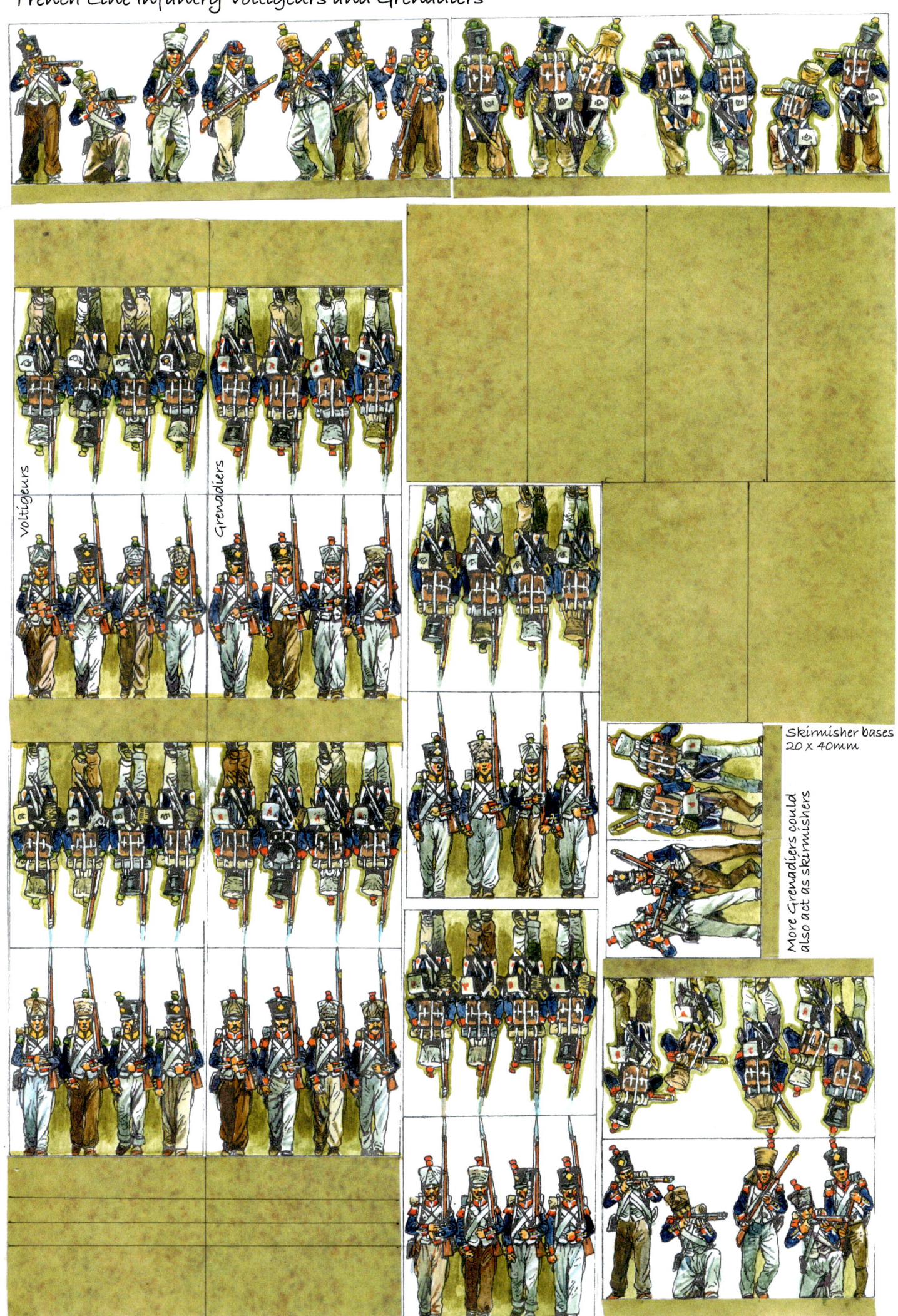

Voltigeurs
Grenadiers
Skirmisher bases 20 x 40mm
More Grenadiers could also act as skirmishers

French line infantry in dustcoats   Note: Command strips on p12
Skirmishers

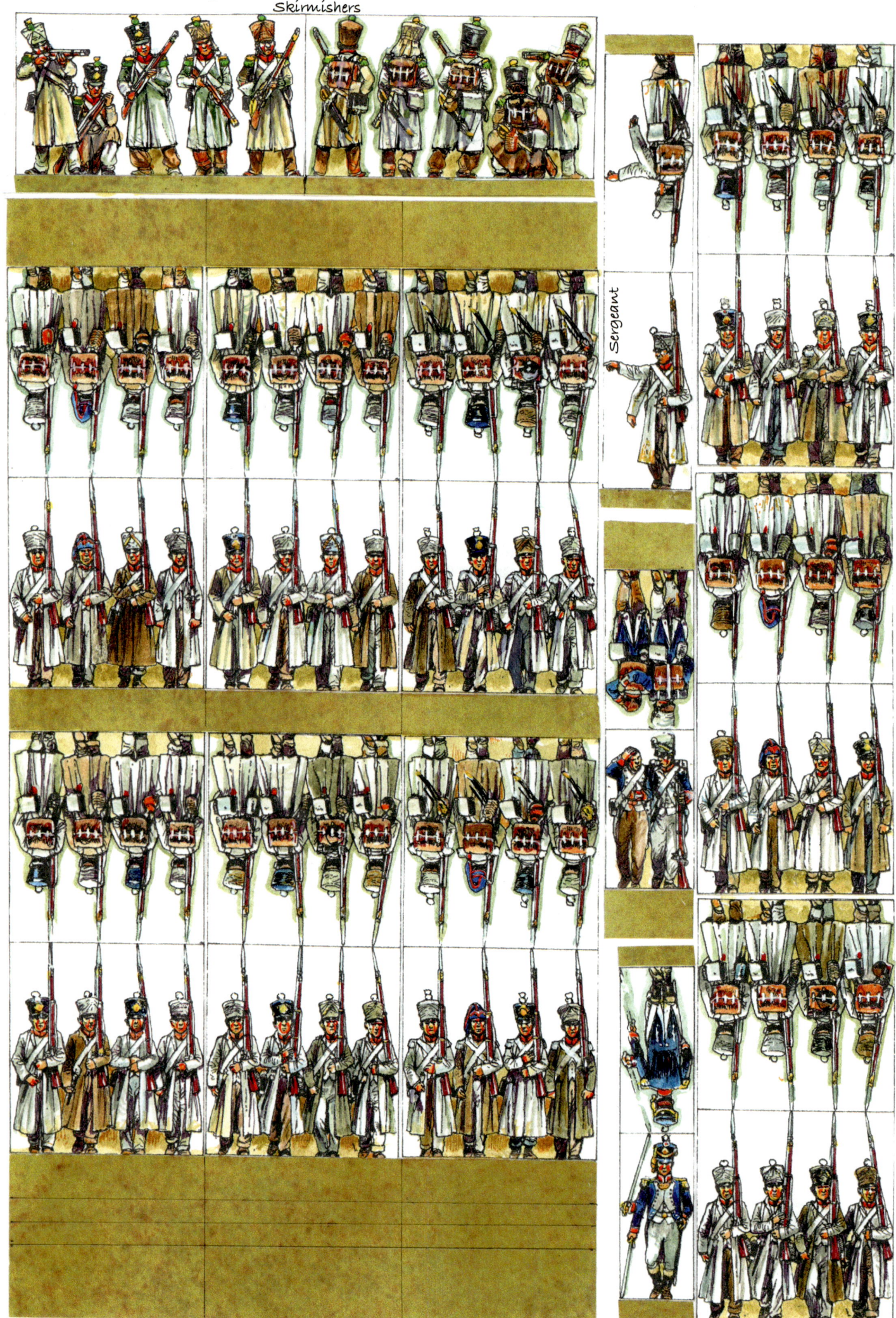

Sergeant

French Light Infantry centre companies

Skirmish bases on Flank Company sheet

Casualty marker

French Light Infantry Voltigeurs and Grenadiers

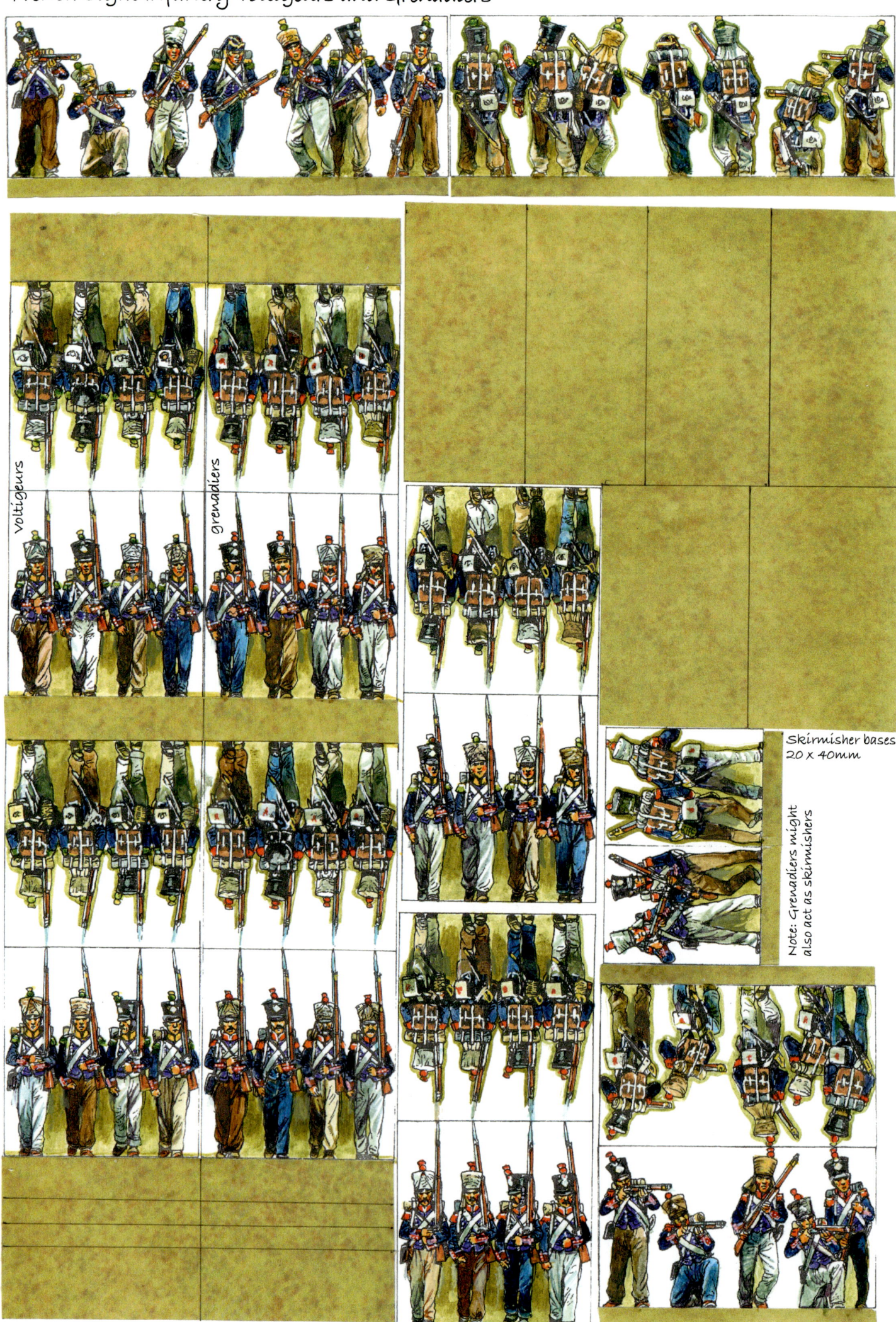

Voltigeurs
Grenadiers

Skirmisher bases
20 x 40mm

Note: Grenadiers might also act as skirmishers

French dismounted Dragoons
use spare flag from mounted Dragoons

Skirmisher bases on inside back cover

Dragoon colour areas. See facings guide page 48 for colours.

Replace with command

Command

Swiss regiment centre companies (French)

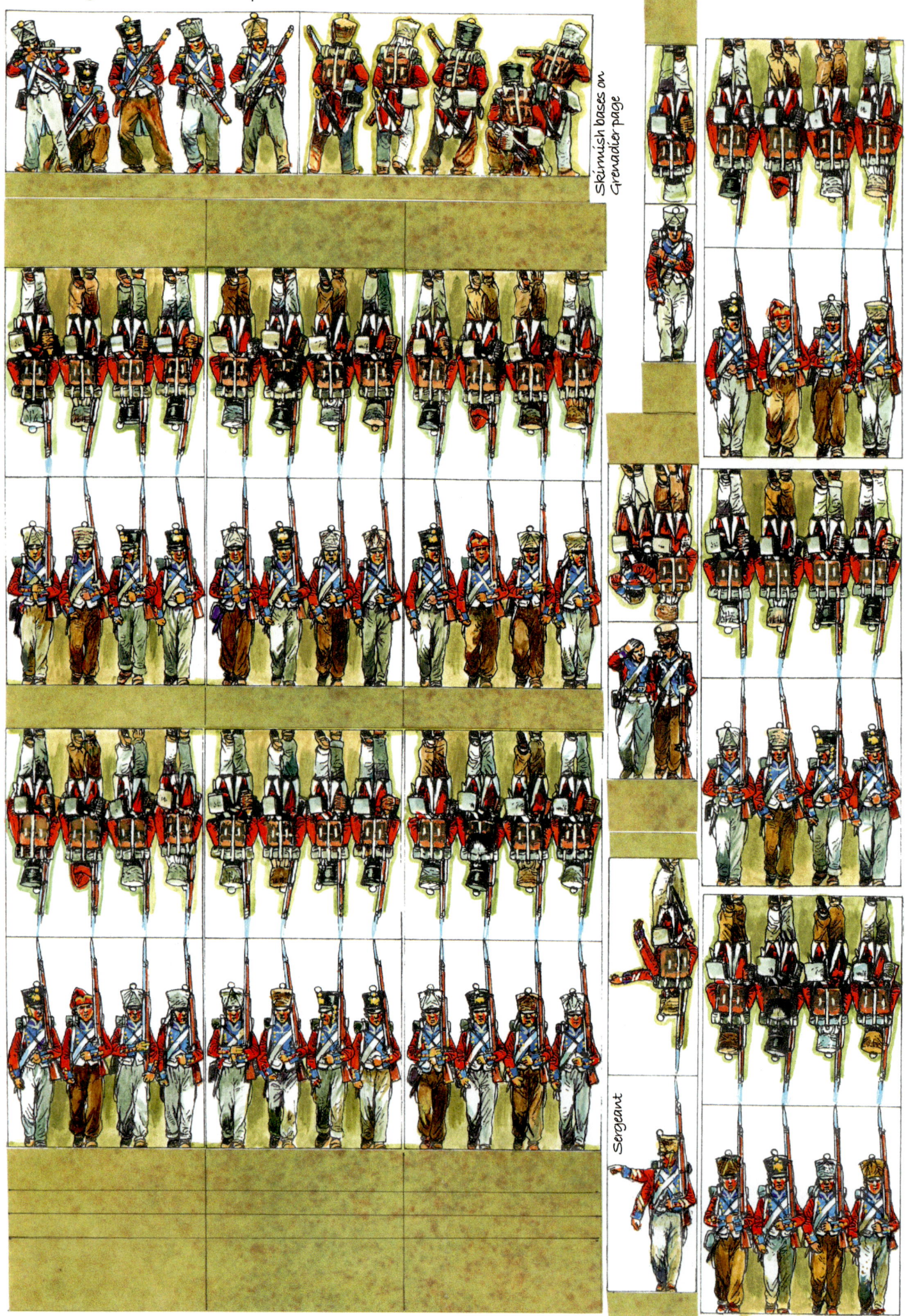

Swiss Regiment Voltigeurs and Grenadiers (French)

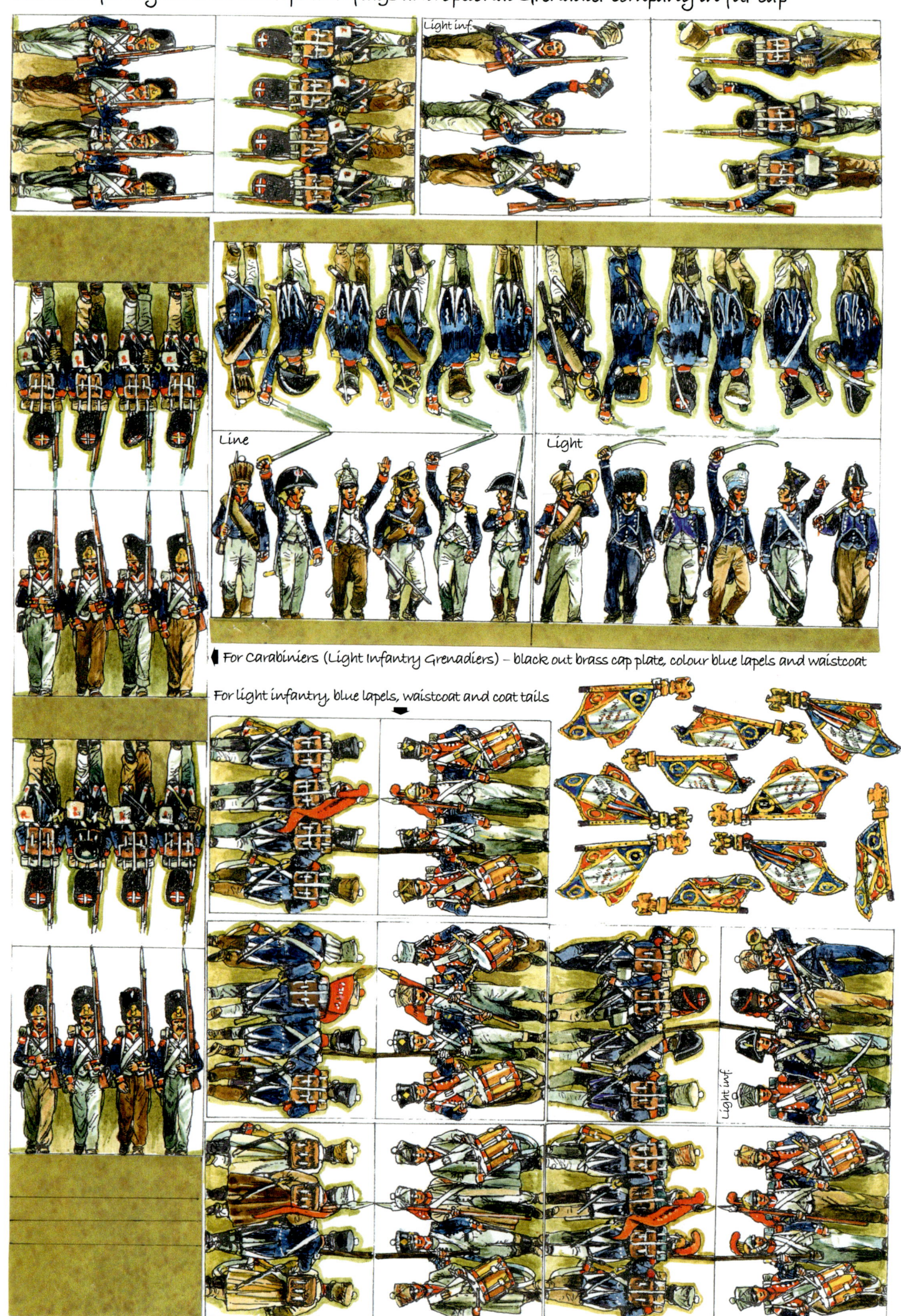

Vistula Legion (Polish troops) (French) infantry

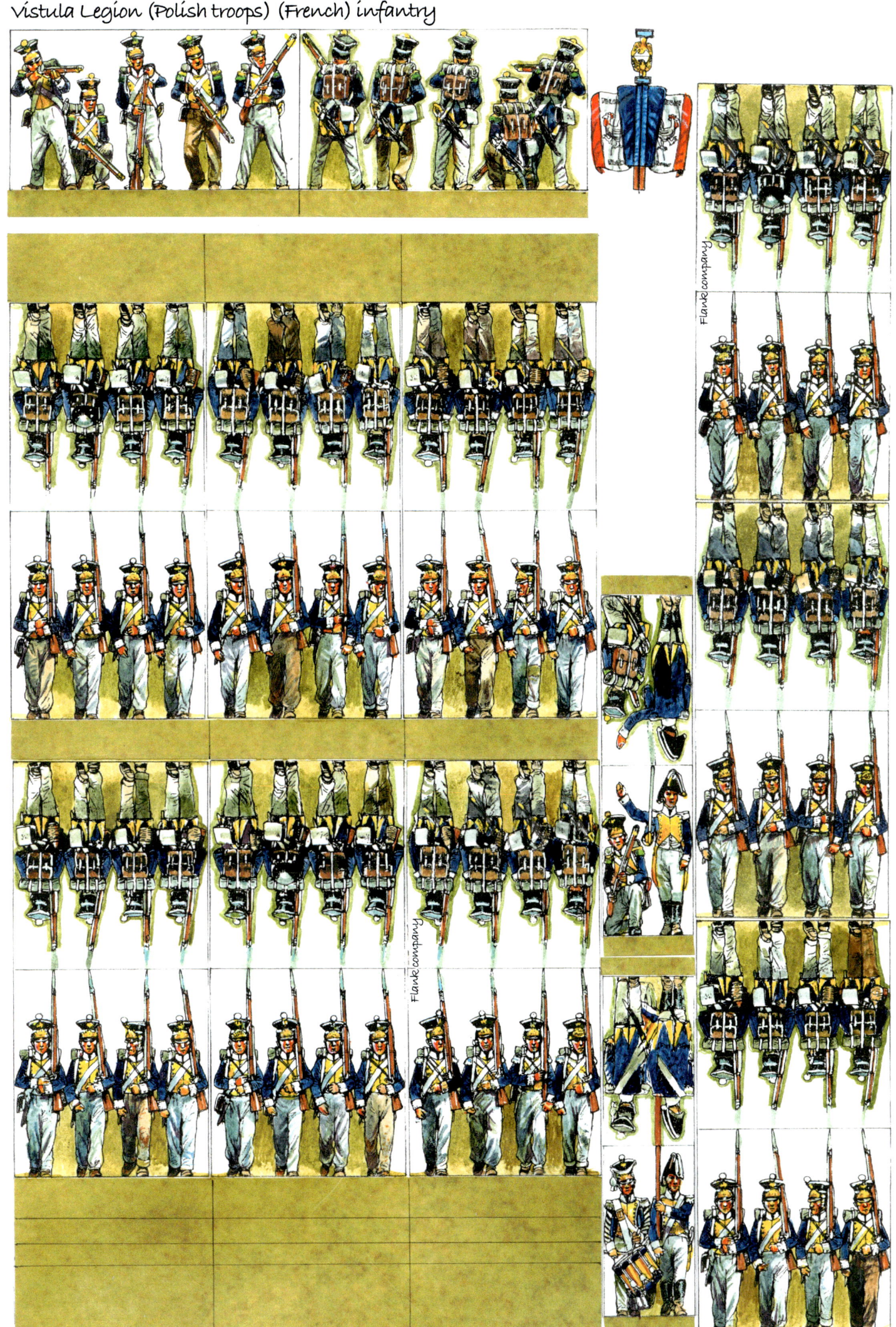

Vistula Legion lancers (French)

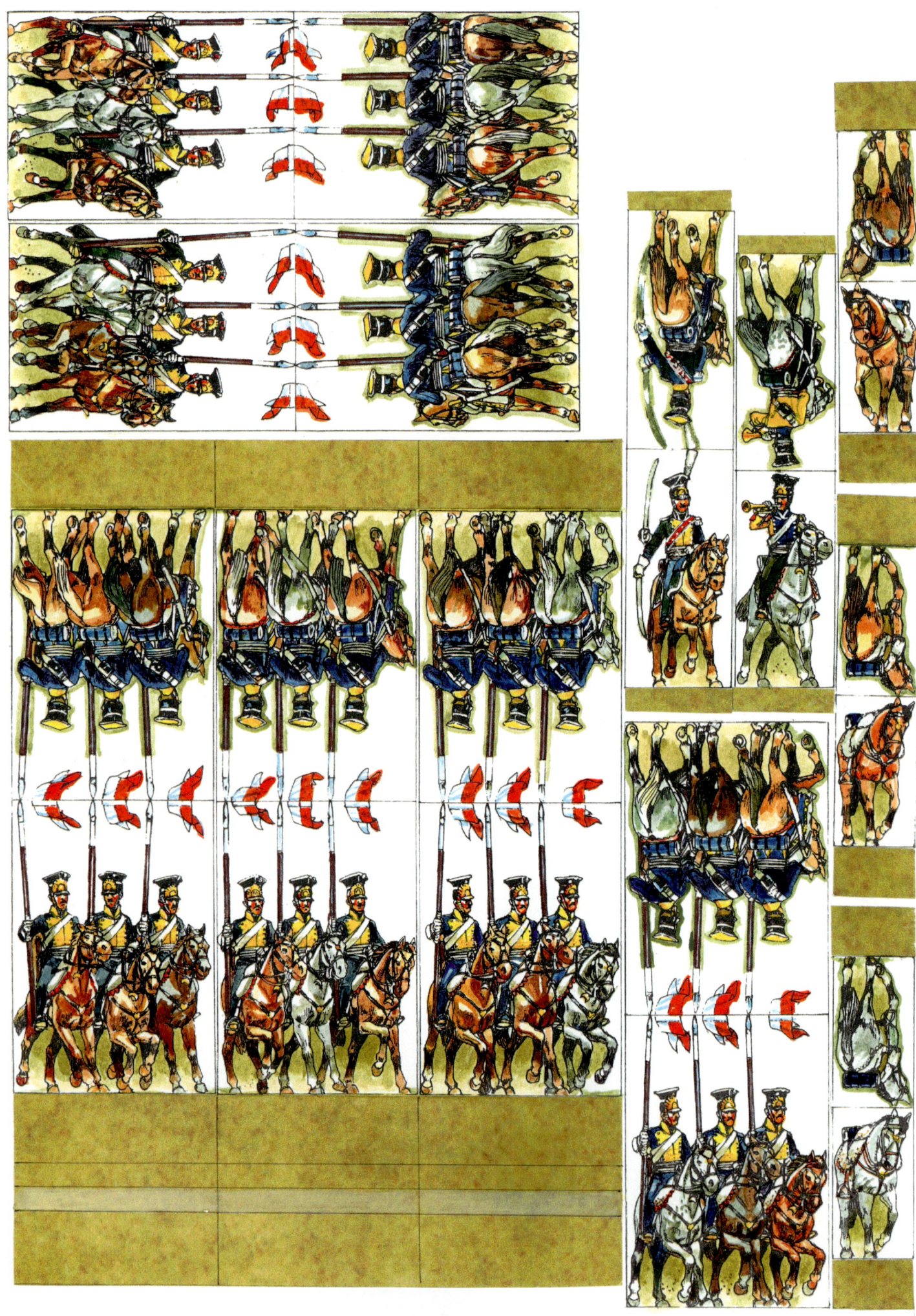

French Dragoons 1 (see 'Facings' guide for colours)

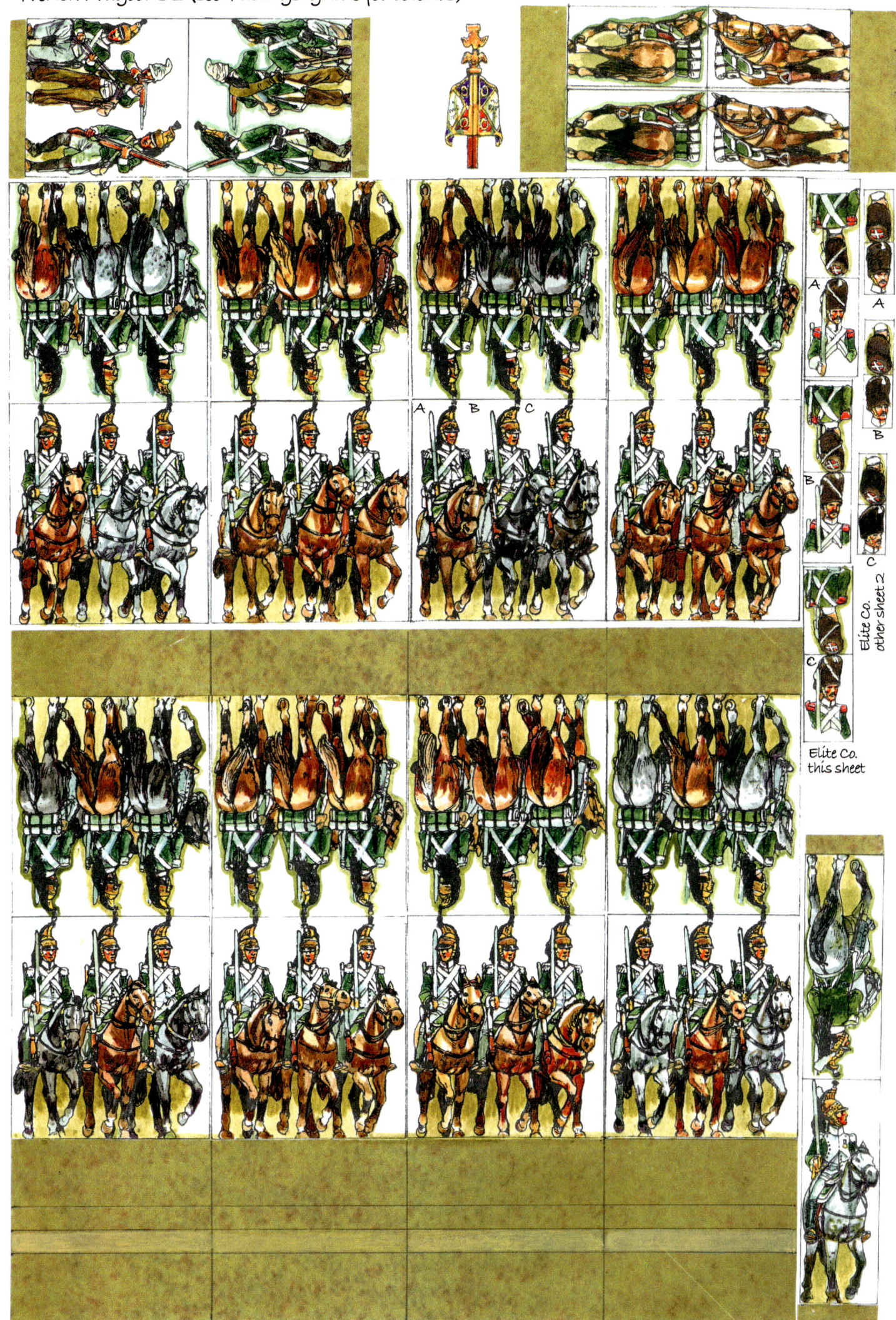

French Dragoons 2 in helmet covers

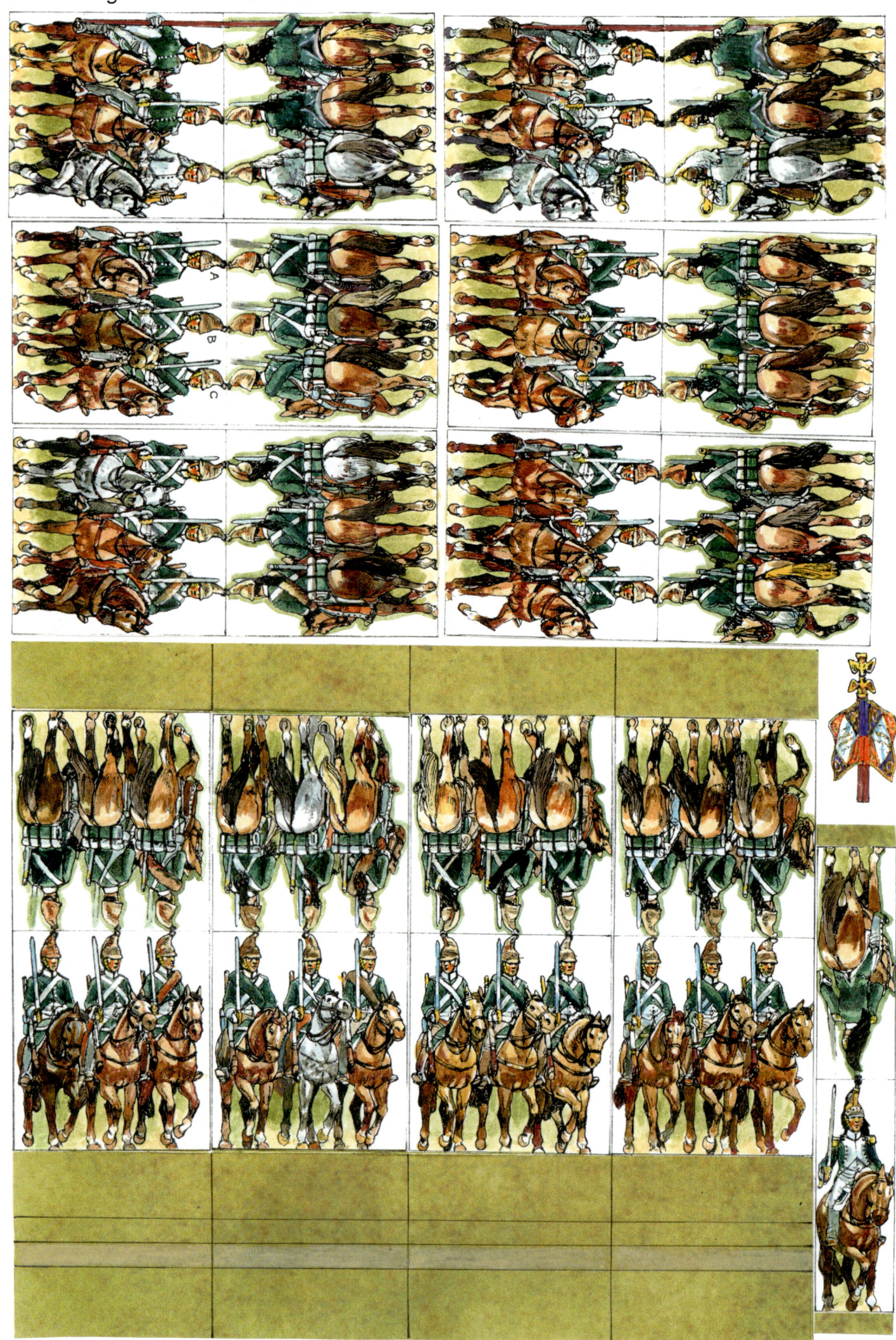

10th Hussars

For 5th Hussars: blue collar, white cuffs and pelisse, yellow braid on chest. For 1st Hussars: blue collar, red sabretache with white edge.

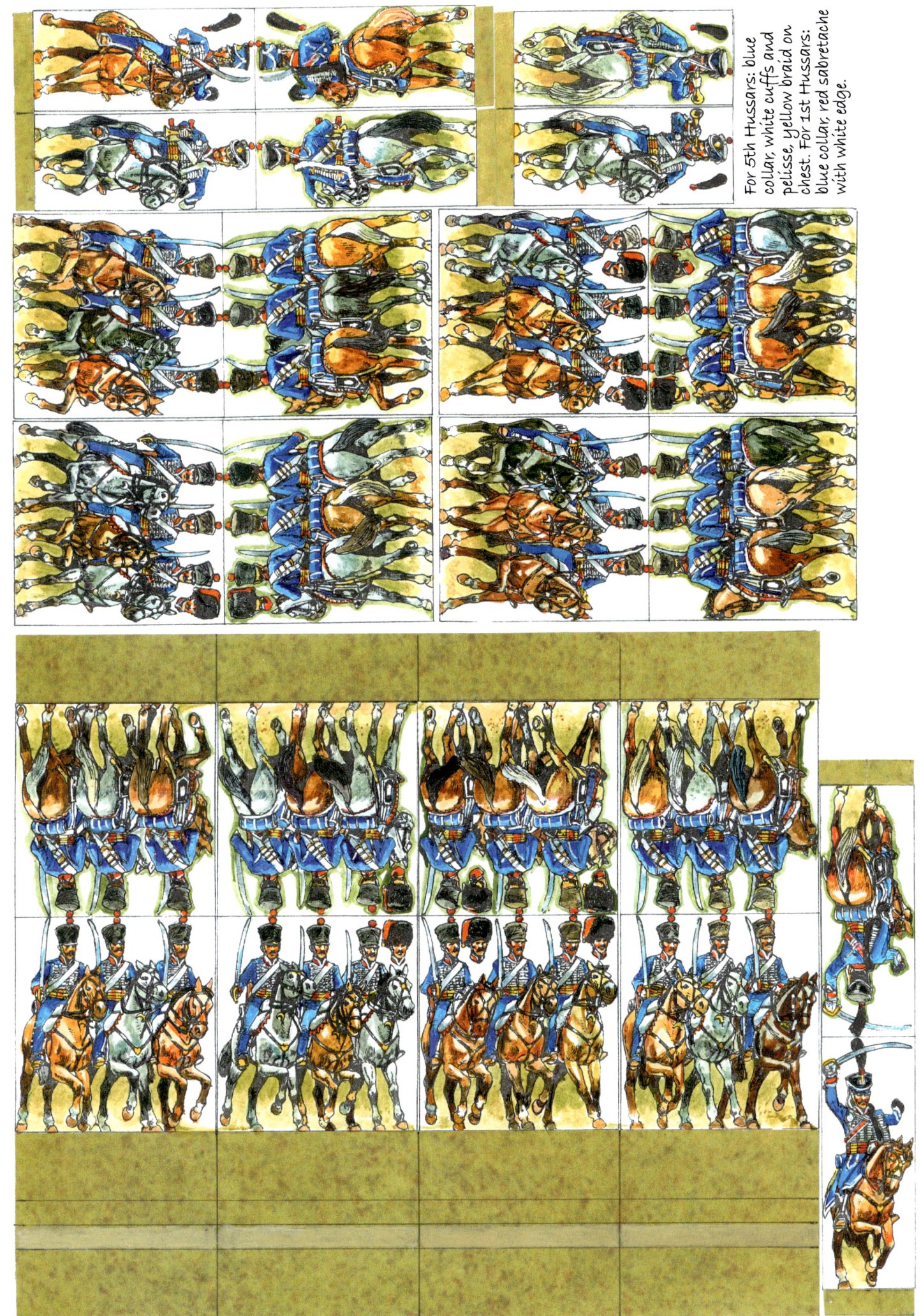

French 12th Hussars

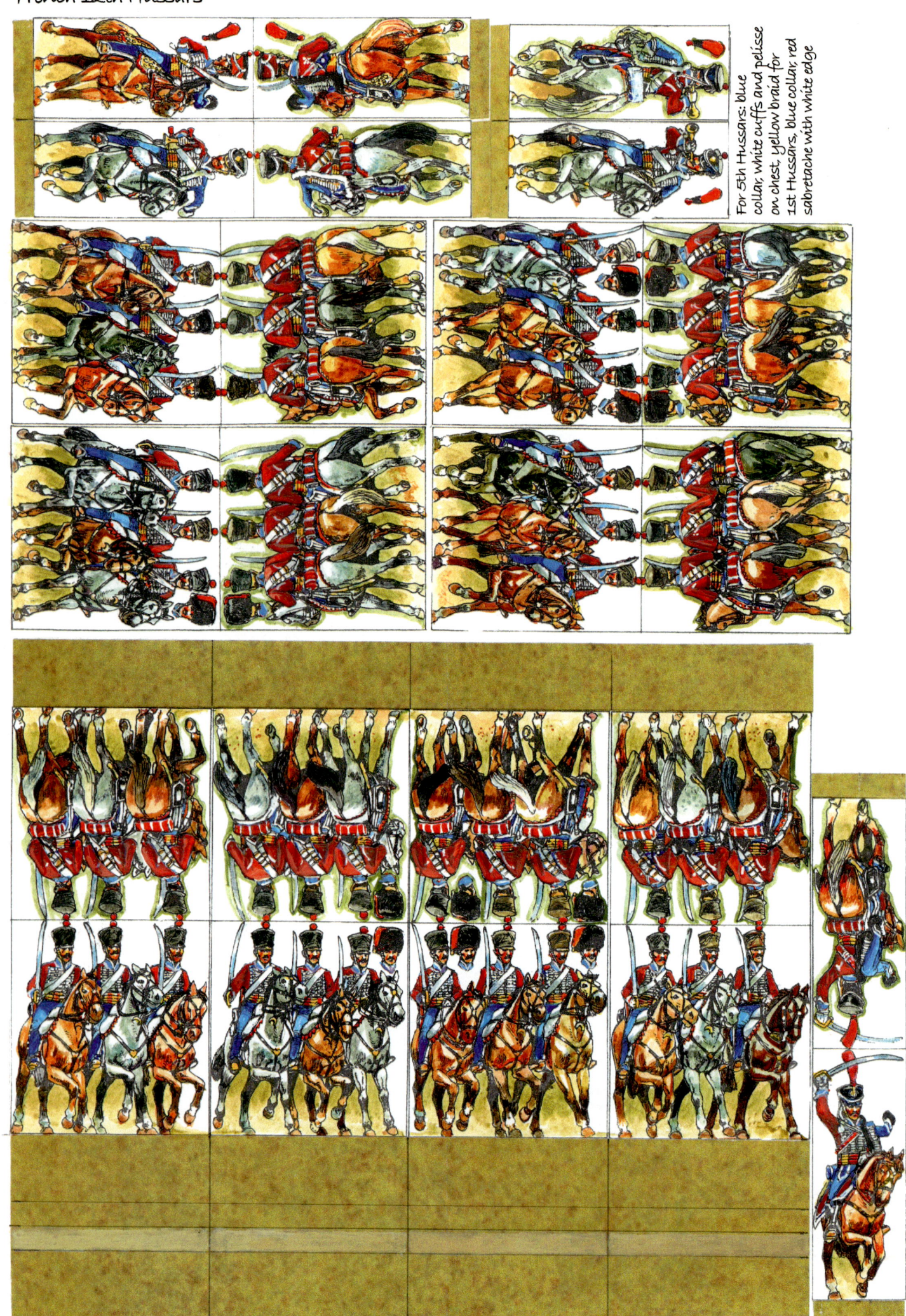

For 5th Hussars: blue collar, white cuffs and pelisse on chest, yellow braid for 1st Hussars, blue collar, red sabretache with white edge

French Chasseurs à Cheval (see 'facings' guide for colours)

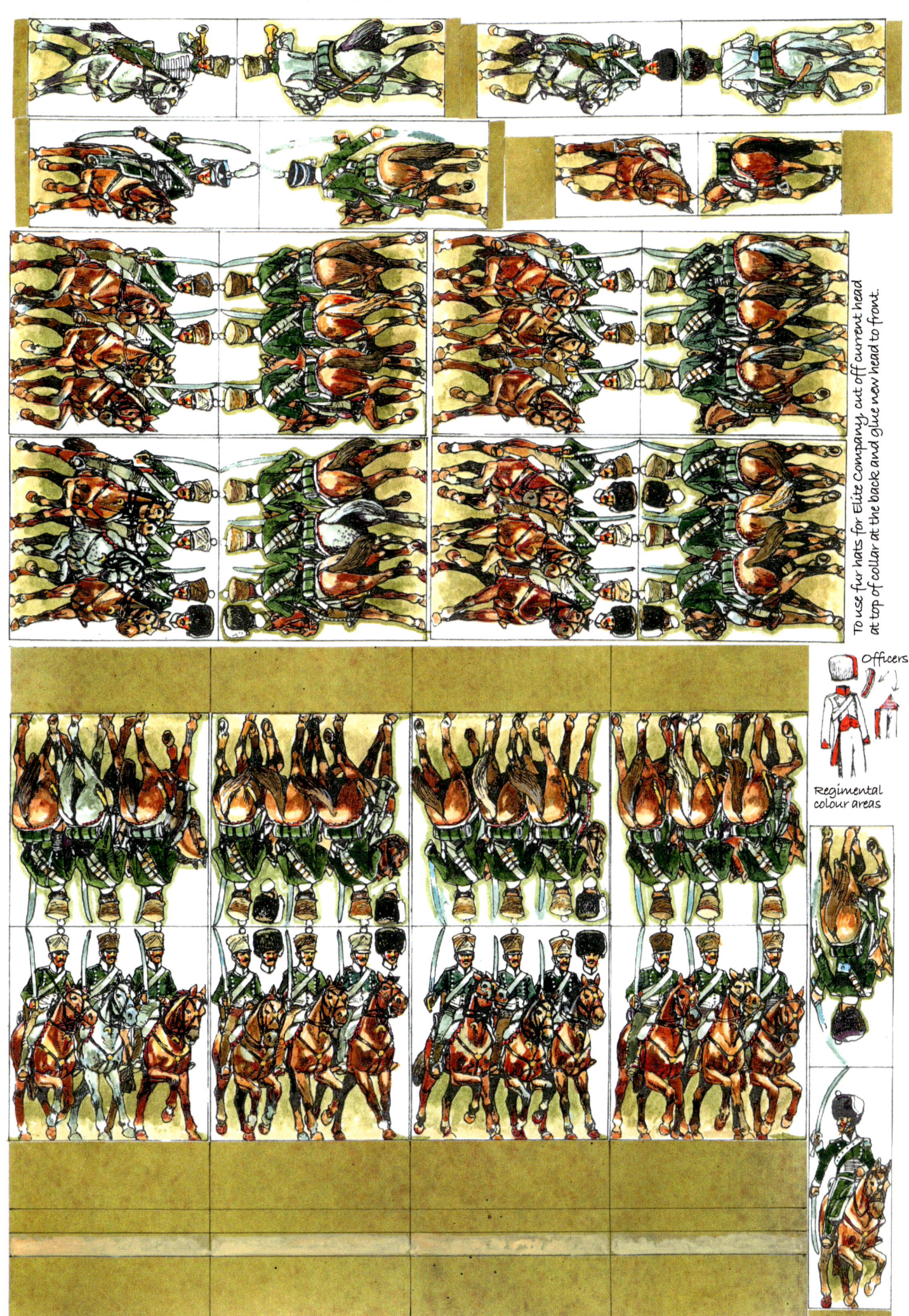

To use fur hats for Elite Company, cut off current head at top of collar at the back and glue new head to front.

Officers

Regimental colour areas

French Artillery (N.B. Extra bases on Helion & Co. website; free download)

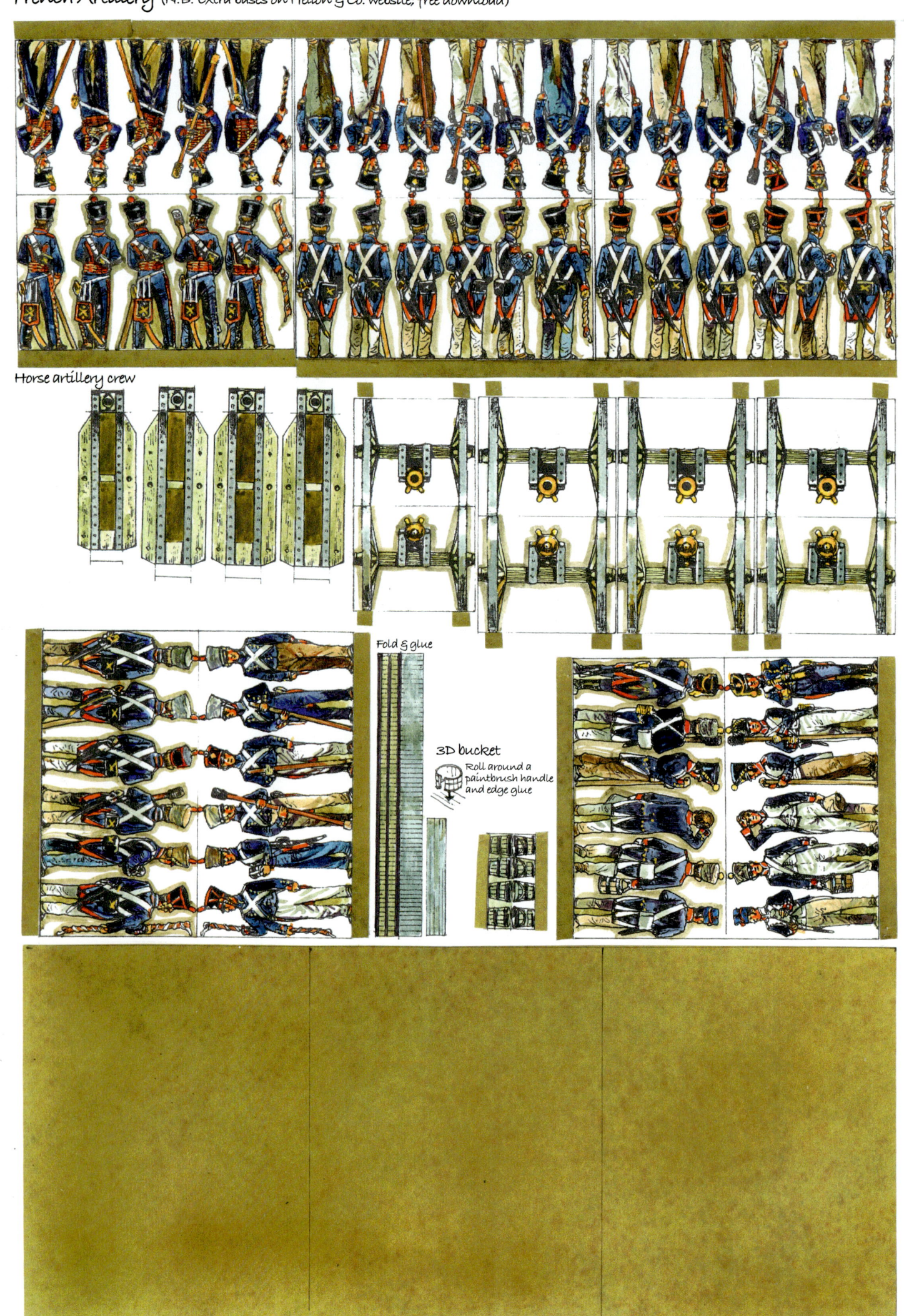

French artillery limbers

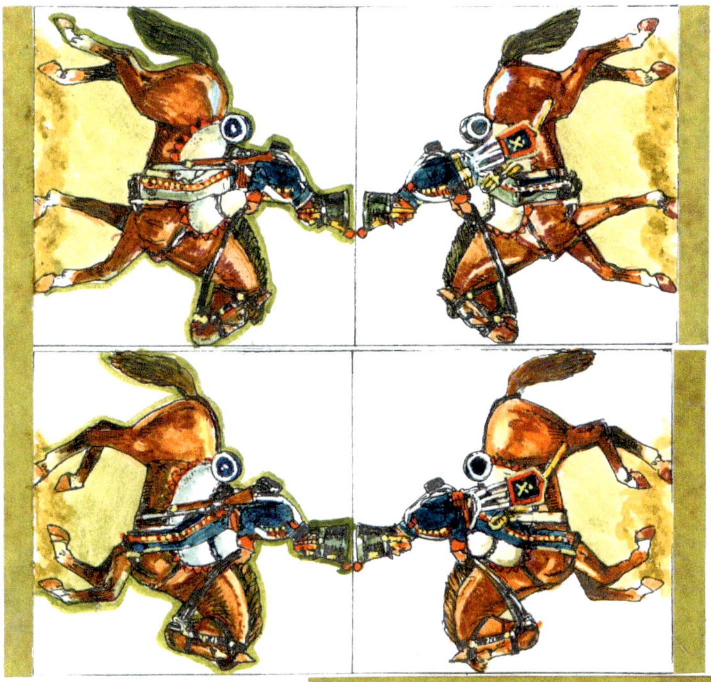

Mounted horse artillery gunners

Note: 12-pounder guns had a longer barrel. Please cut off the 'swell' of the short barrel when making limbers for 12-pounders.

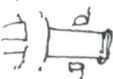

cross bar

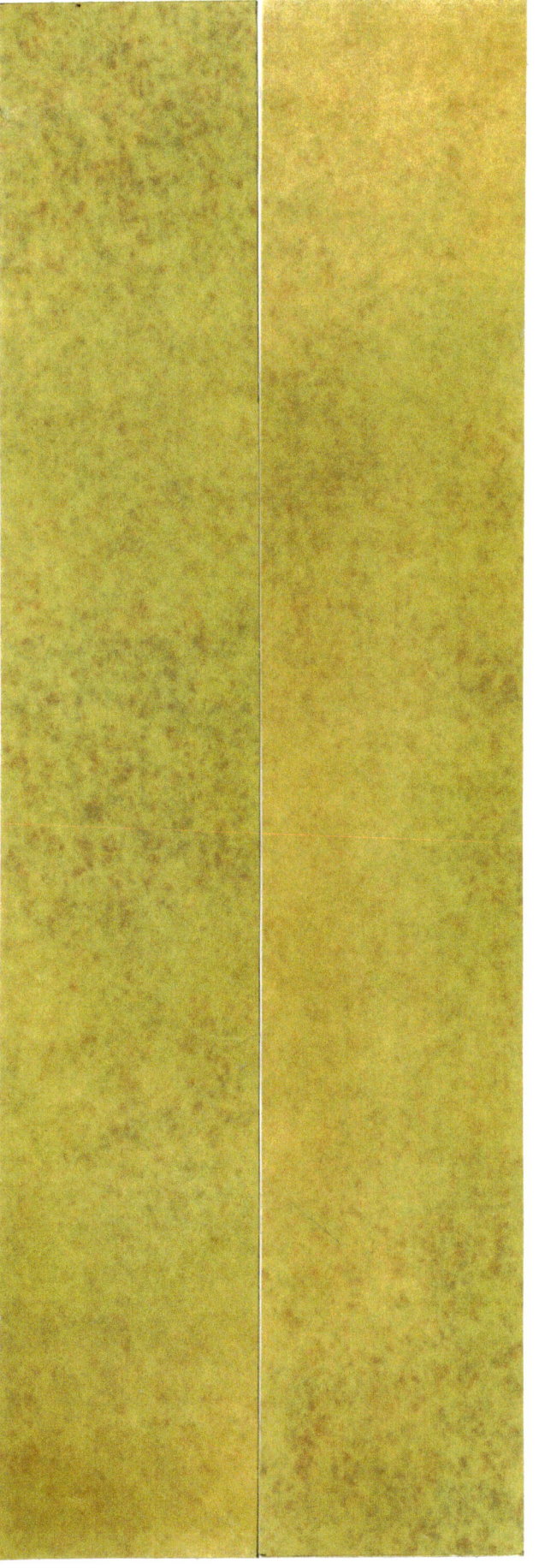

French artillery teams

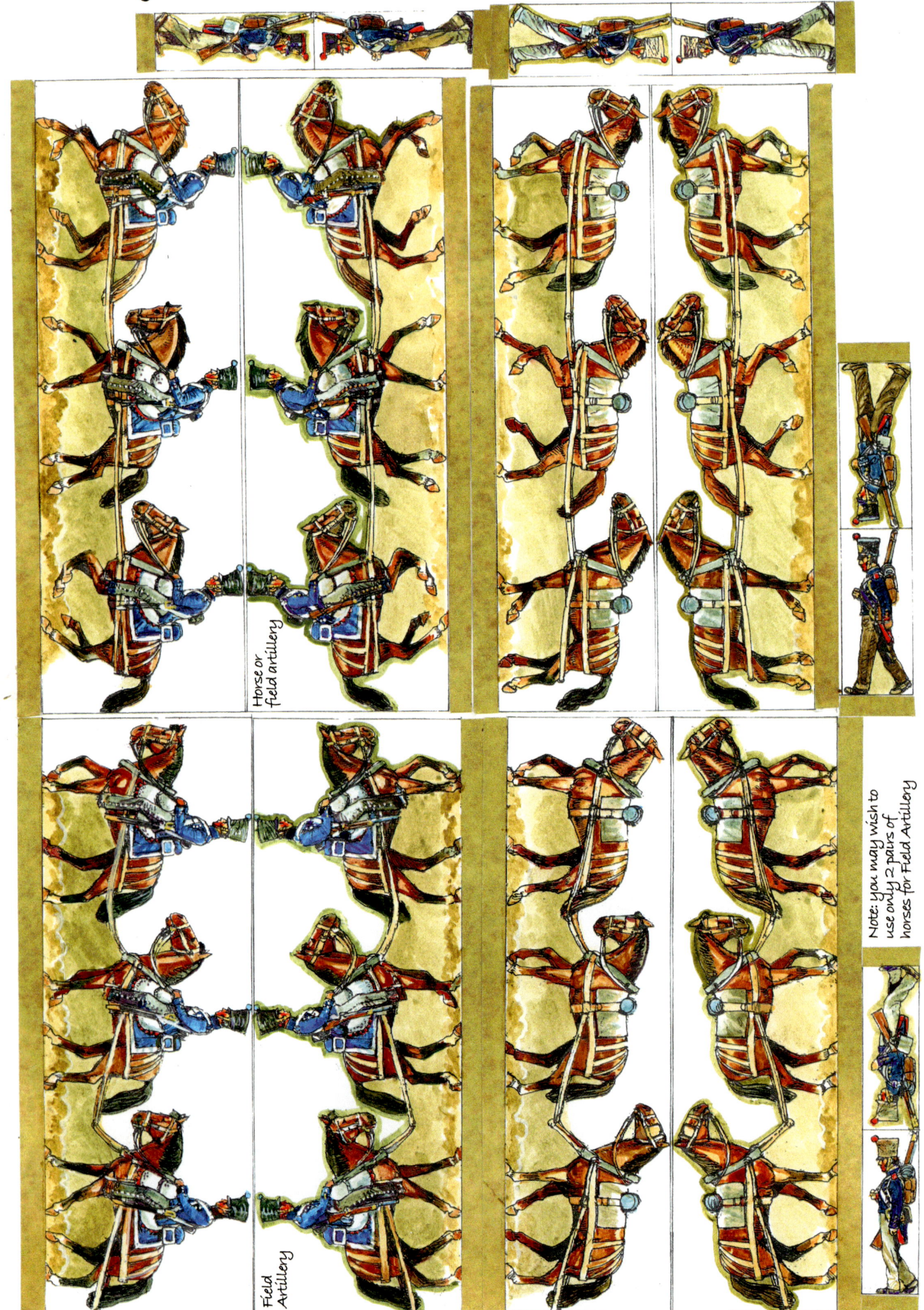

British infantry centre companies, white facings

Light Infantry Regiment
Front rank

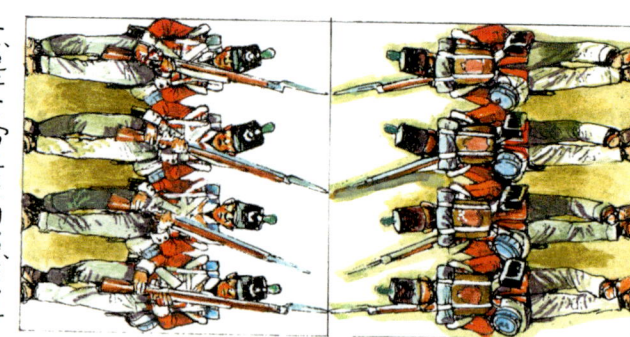

British infantry centre companies, yellow facings

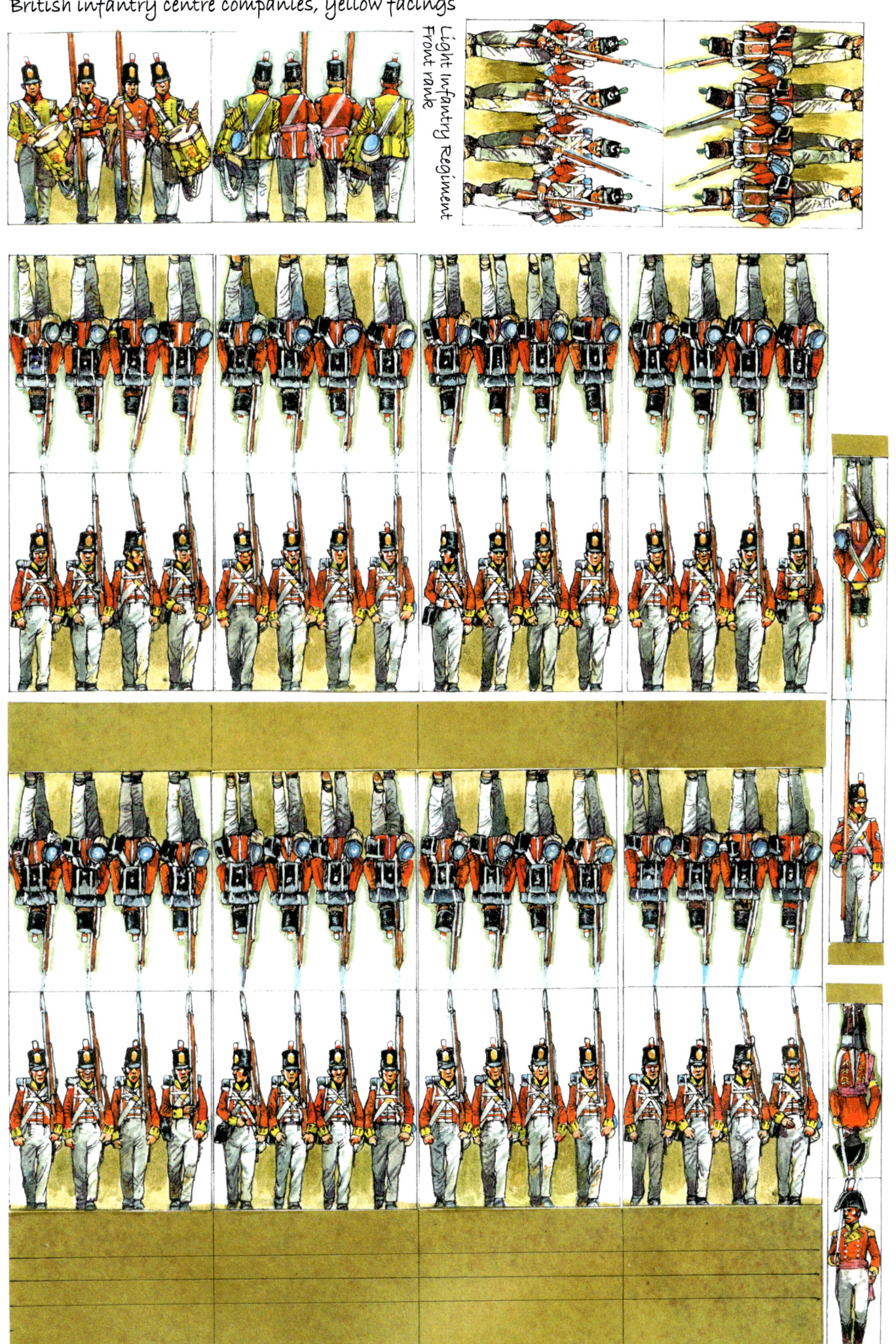

British Infantry centre companies, Guards and Royal Regiments, blue facings. Also King's German Legion.

Light infantry regiment Front rank

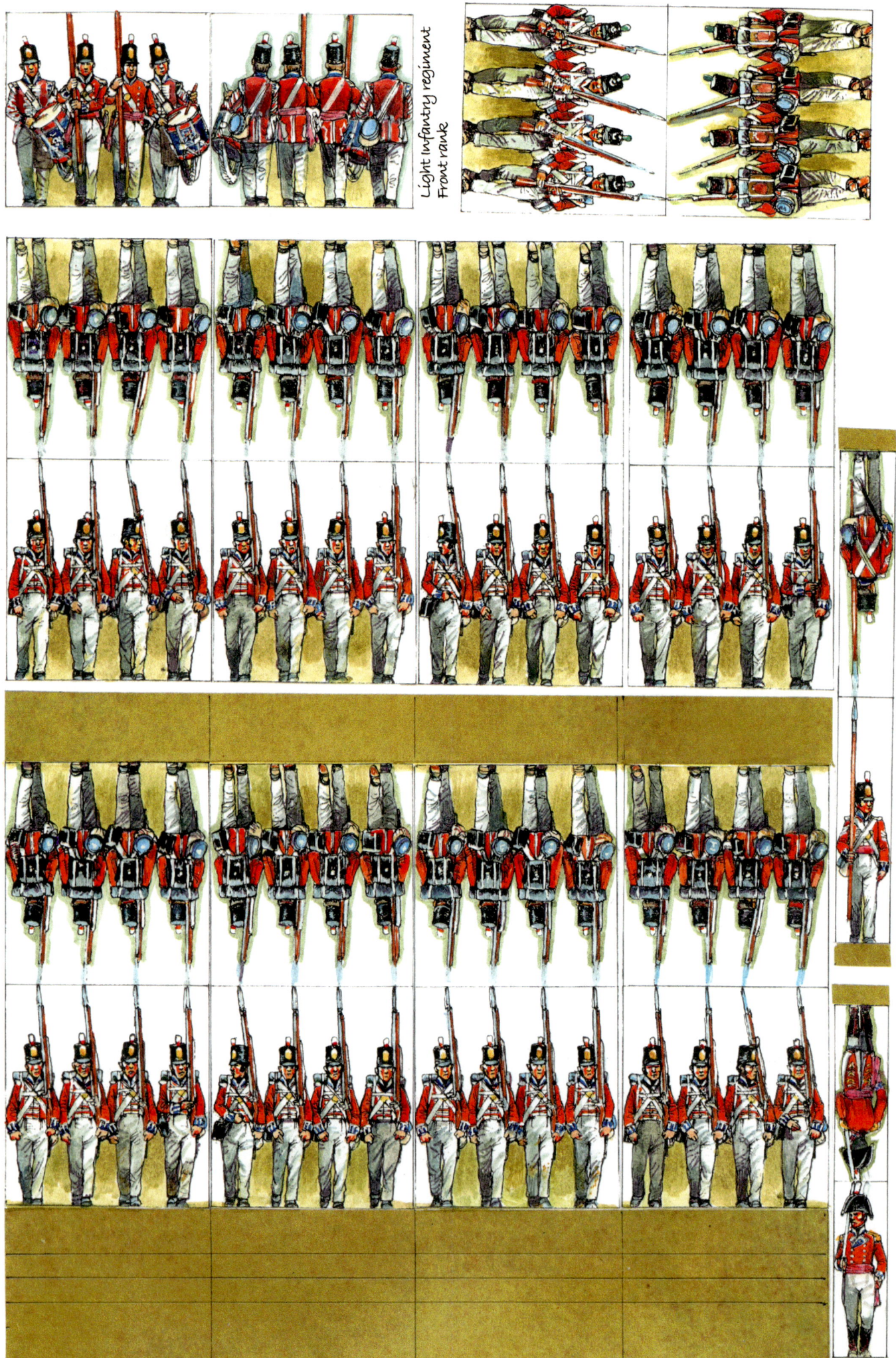

Flank companies and Light Infantry regiments (see 'facings' guide for colours)

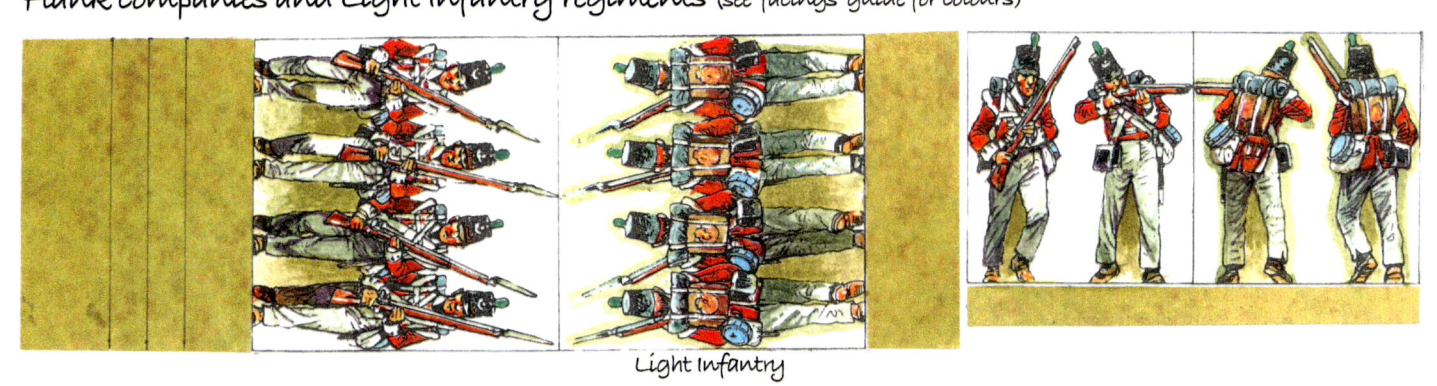

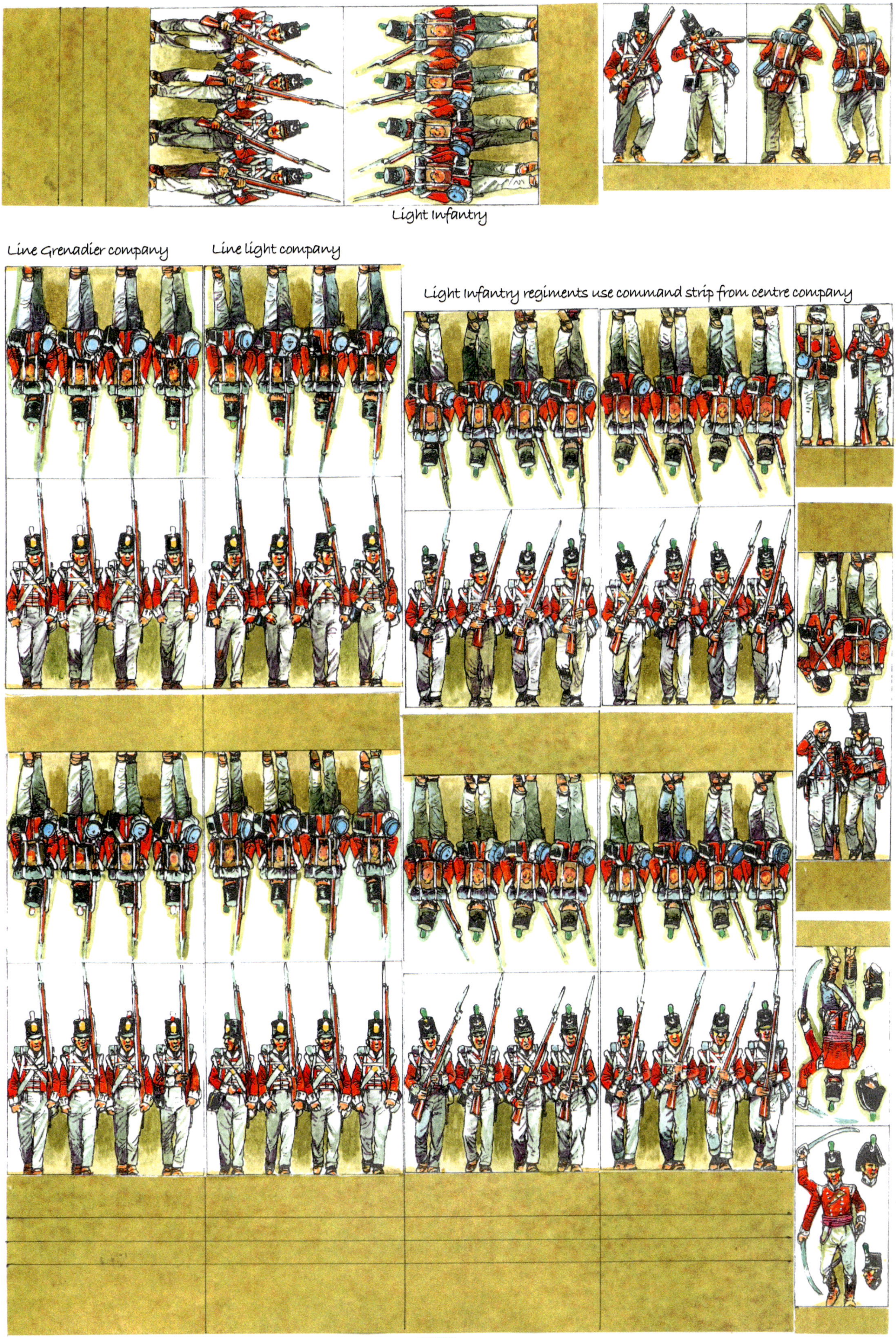

Light Infantry

Line Grenadier company Line light company

Light Infantry regiments use command strip from centre company

British Light Infantry and riflemen in skirmish order

Rifles

Bases 40mm x 20mm

British, Portuguese and Spanish infantry flags

British: one king's colour, one facing colour flag per battalion.
Portuguese: regiments 1, 4, 7, 10, 13, 16, 19, 22 white; 2, 5, 8, 11, 14, 17, 20, 23 red; 3, 6, 9, 12, 15, 18, 21, 24 yellow

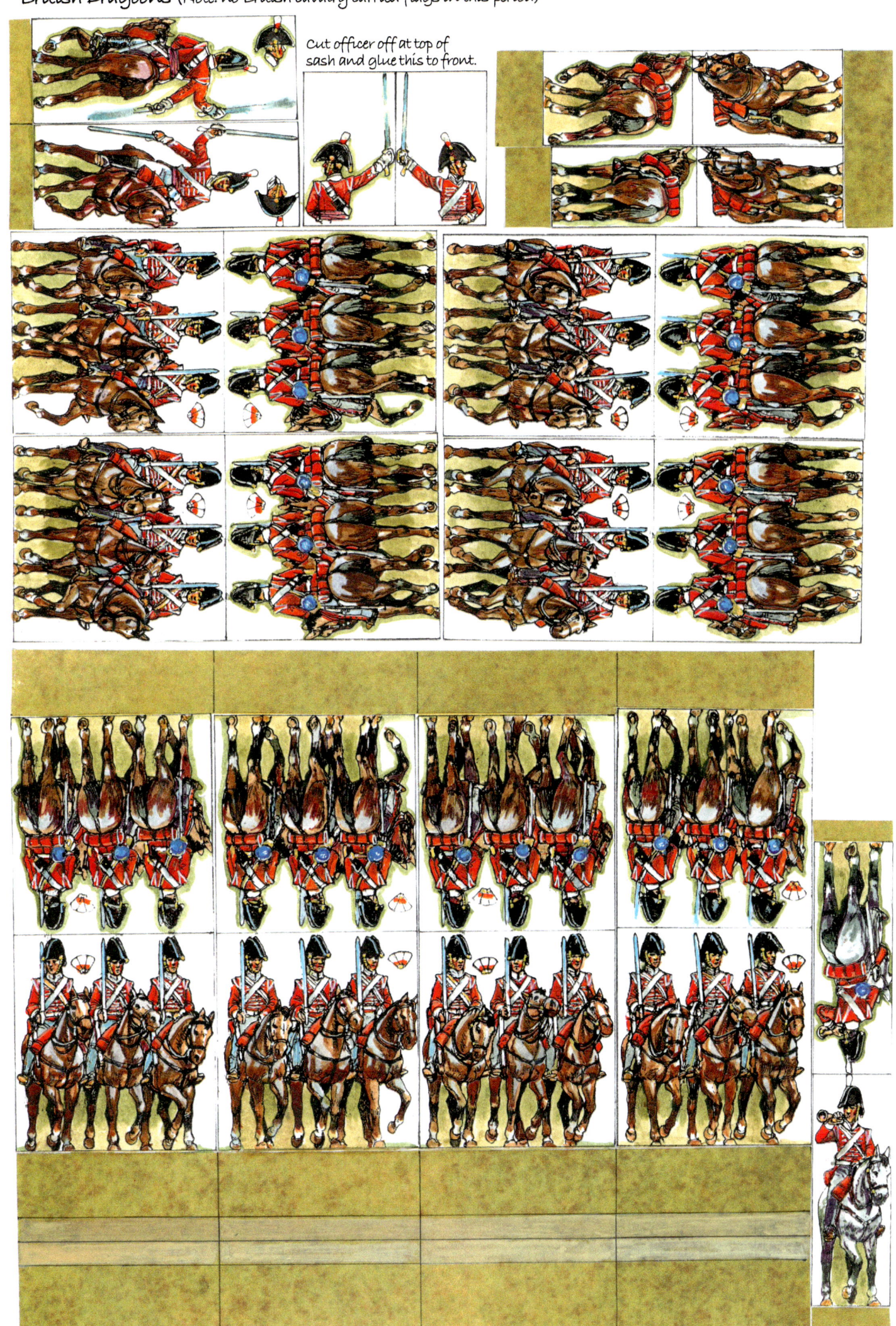

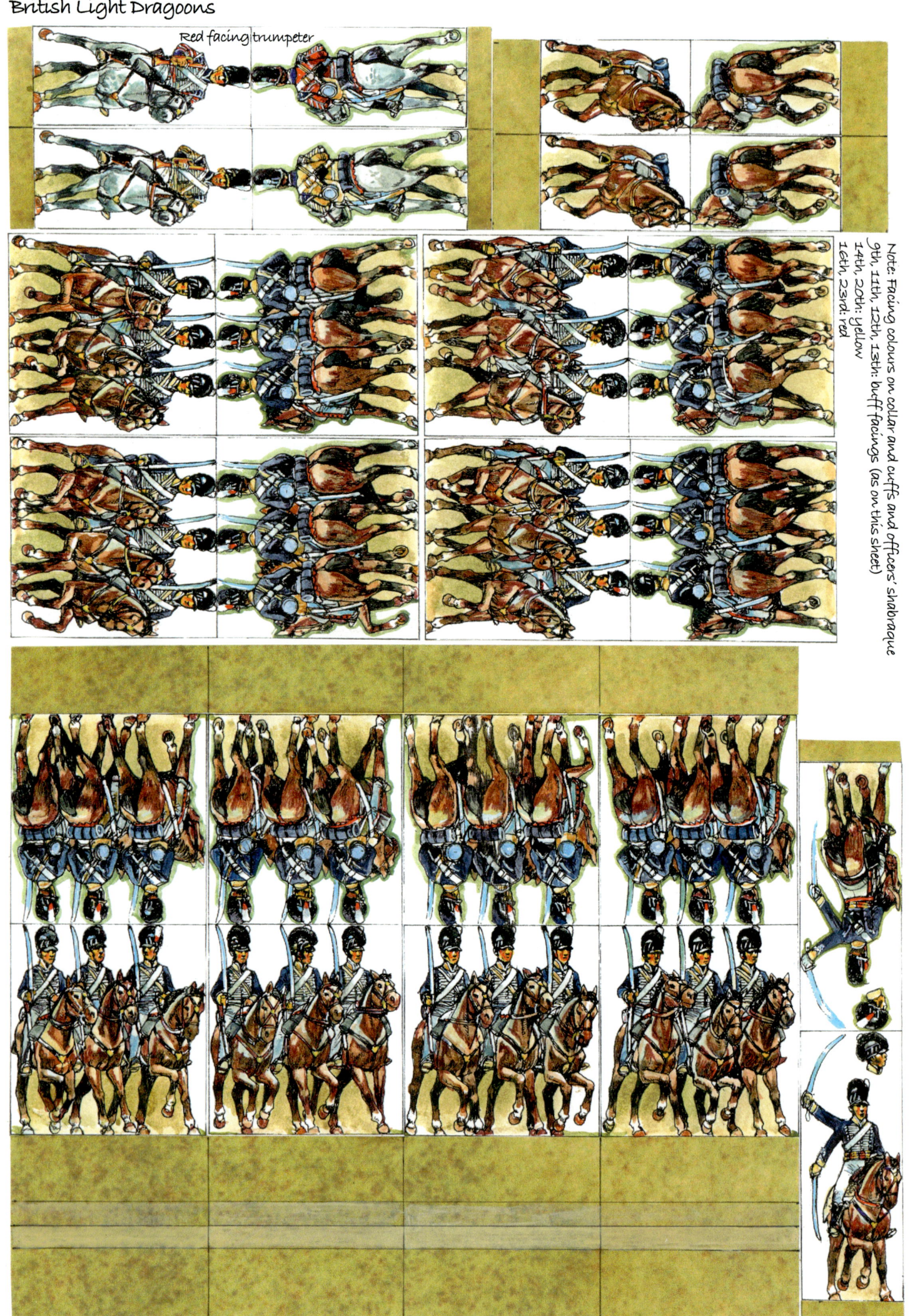

British Hussars

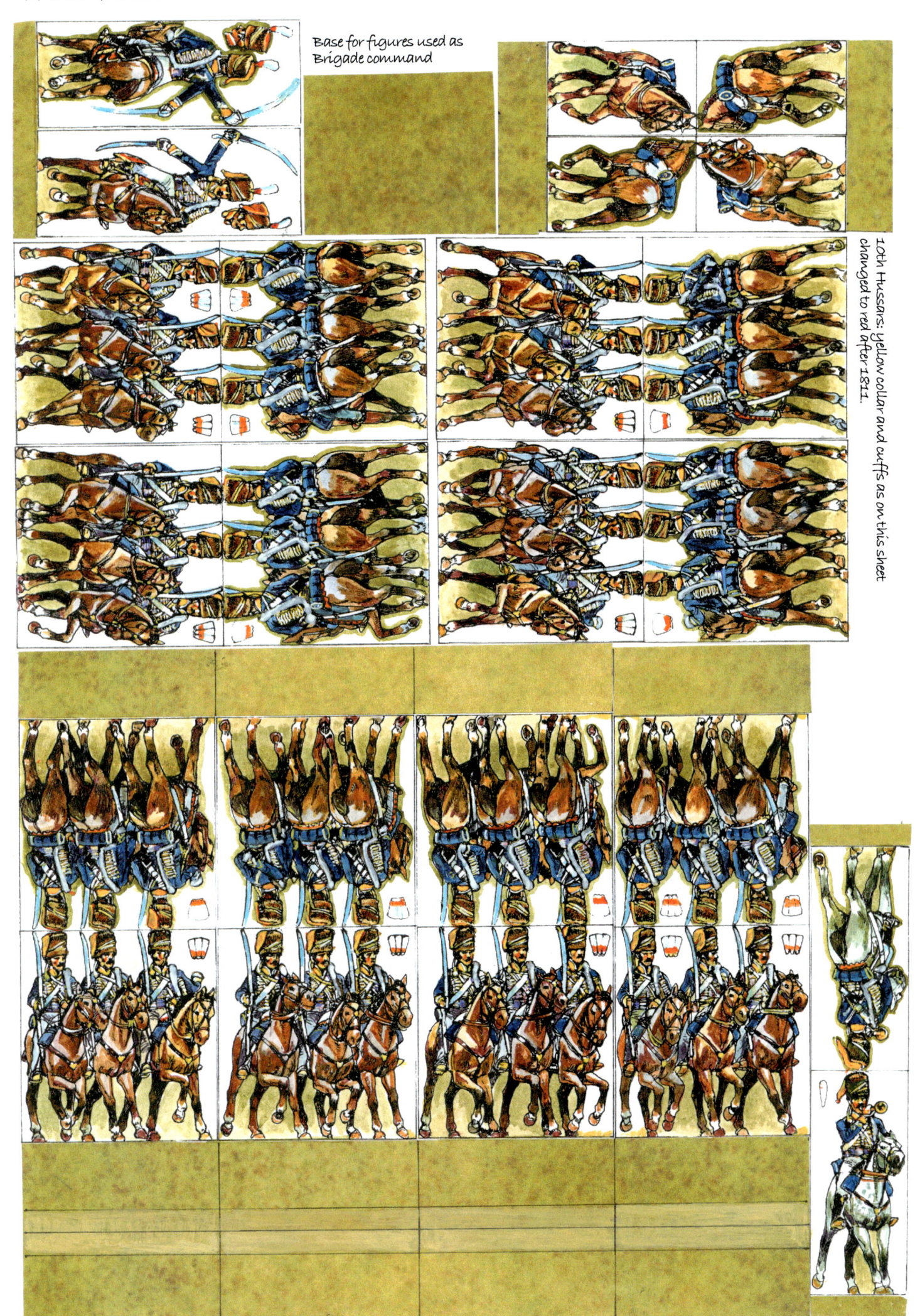

Base for figures used as Brigade command

10th Hussars: yellow collar and cuffs as on this sheet changed to red after 1811.

British and Portuguese Artillery (Note: extra bases on Helion & Co website, free download)

Leather buckets. Fold, then roll and glue to base.

Port.

R.H.A.

British Artillery Limbers

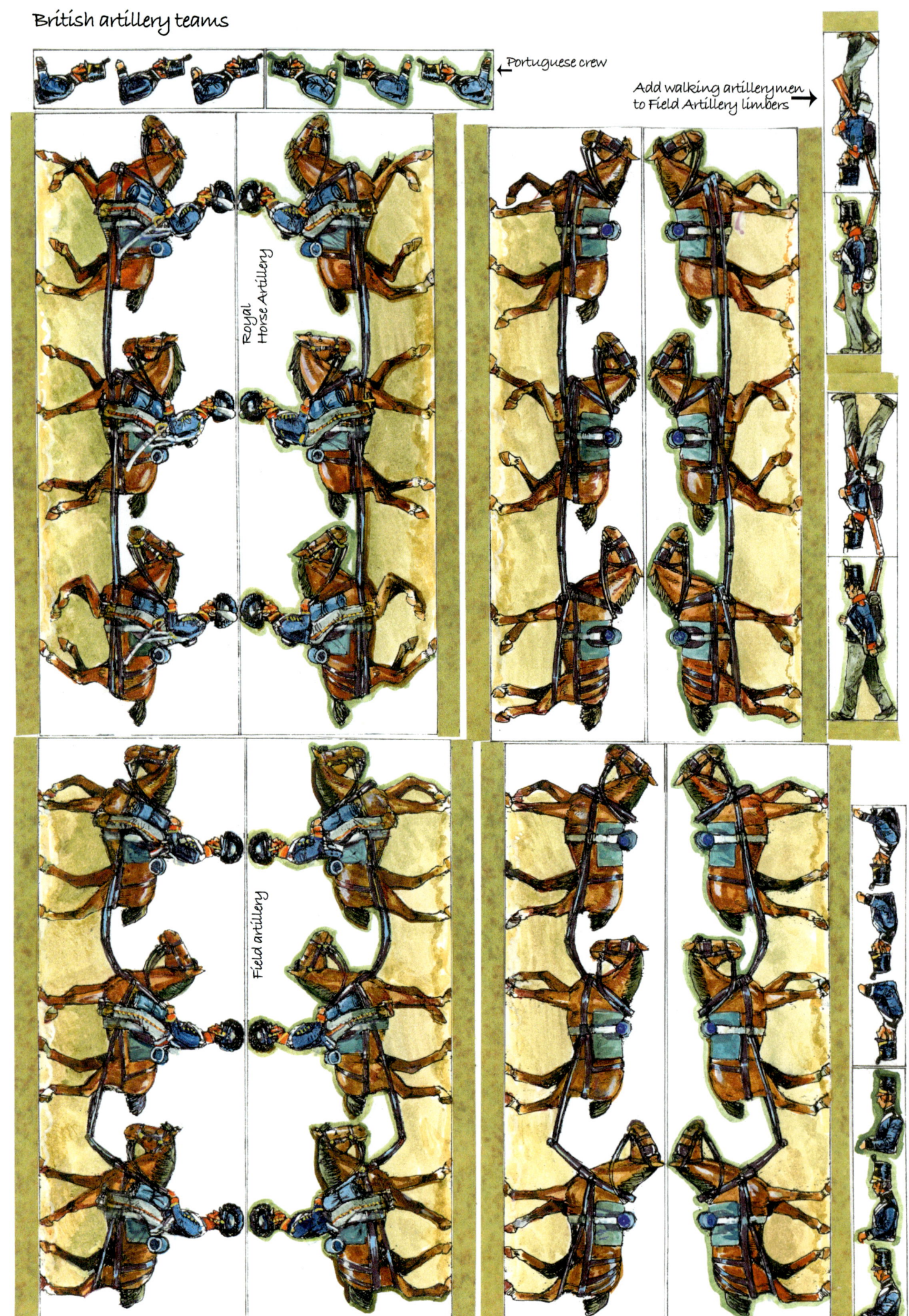

Portuguese Loyal Lusitanian Legion

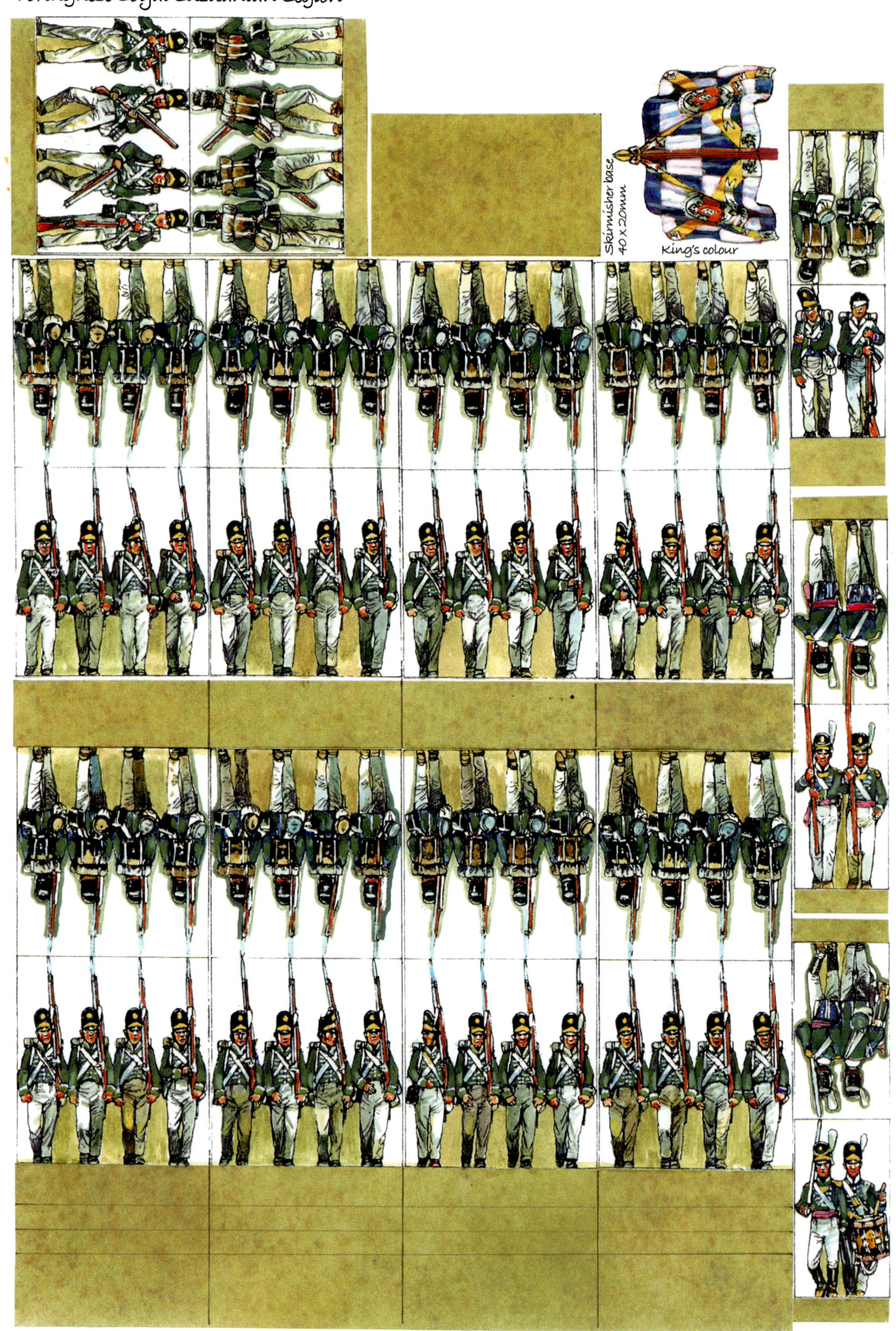

Portuguese Caçadores Light Infantry

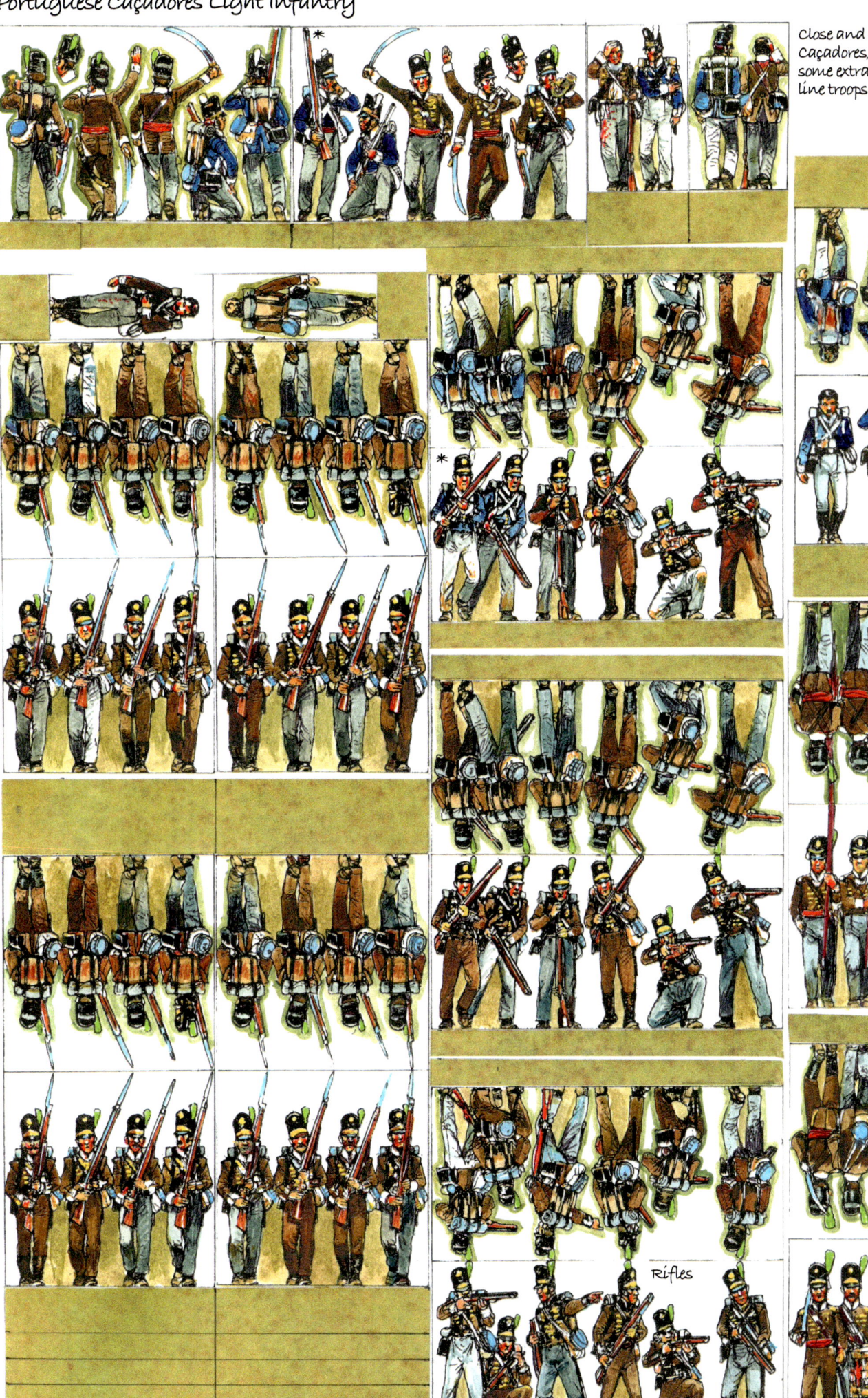

Close and open order Caçadores, including some extra skirmisher line troops marked *

Rifles

Use the skirmish stands on page 8

Portuguese Cavalry

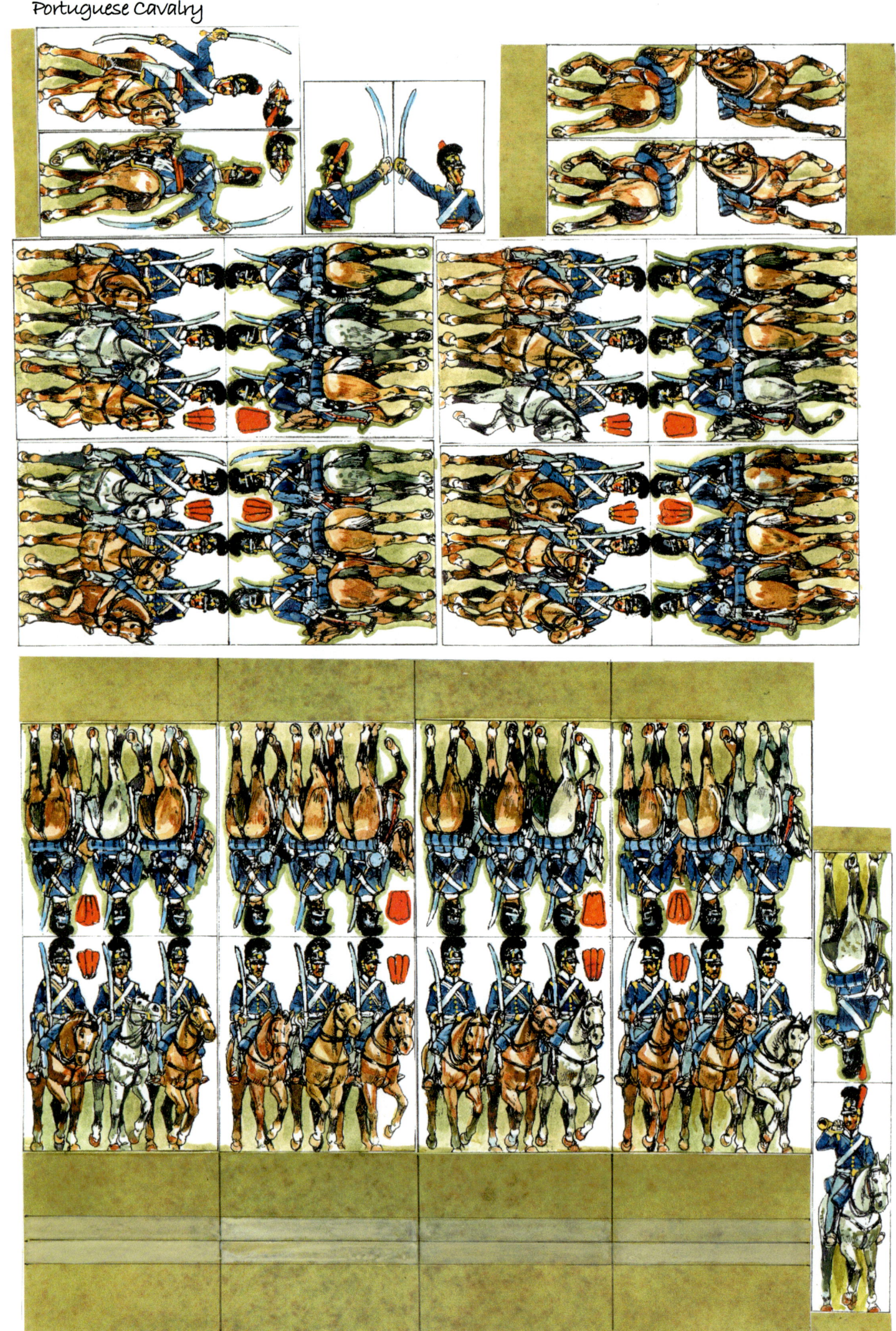

Spanish Infantry, blue coats

Skirmish stands on page 47

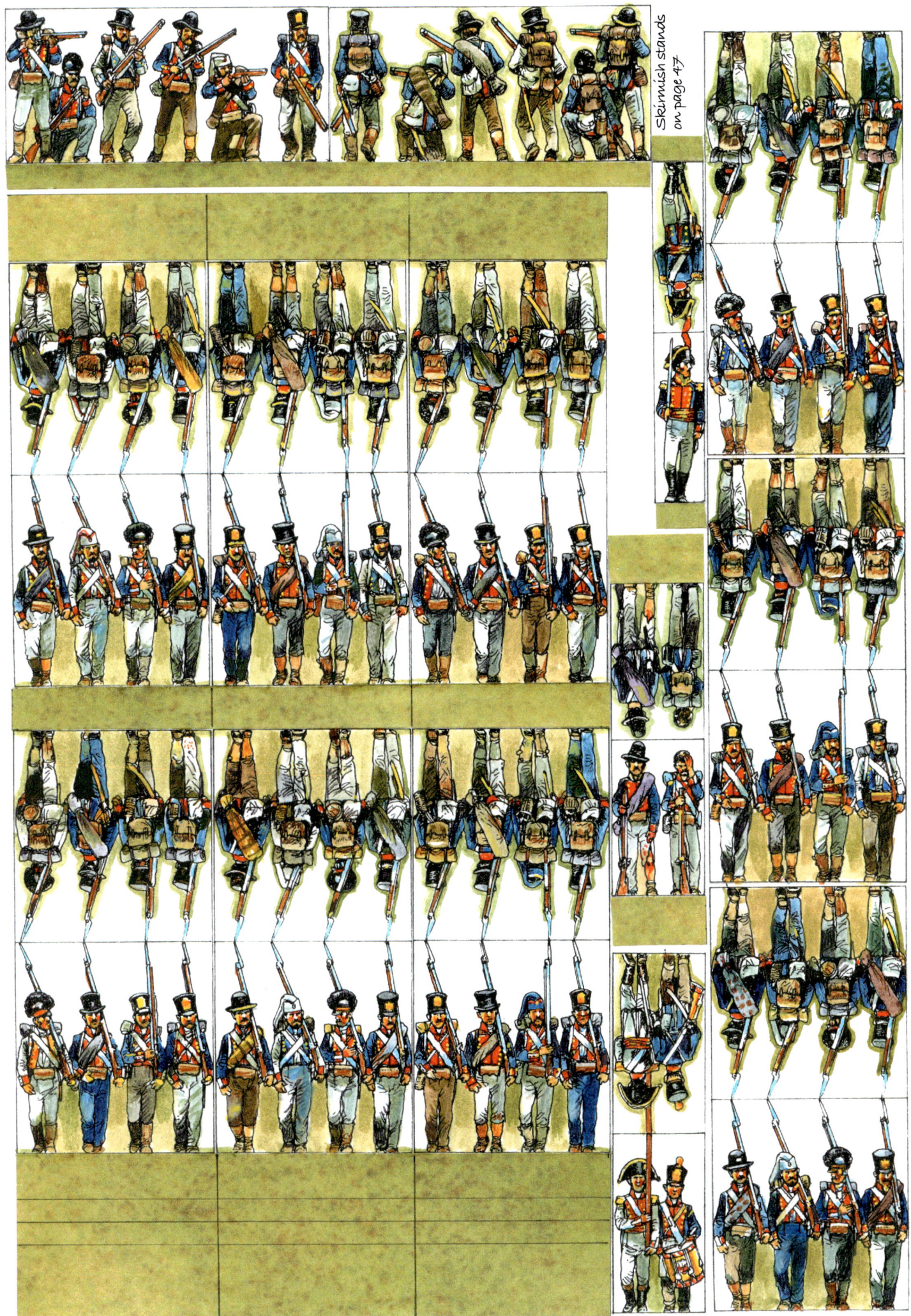

Spanish Line Infantry, brown coats

Skirmish stands on page 47

Spanish Cavalry

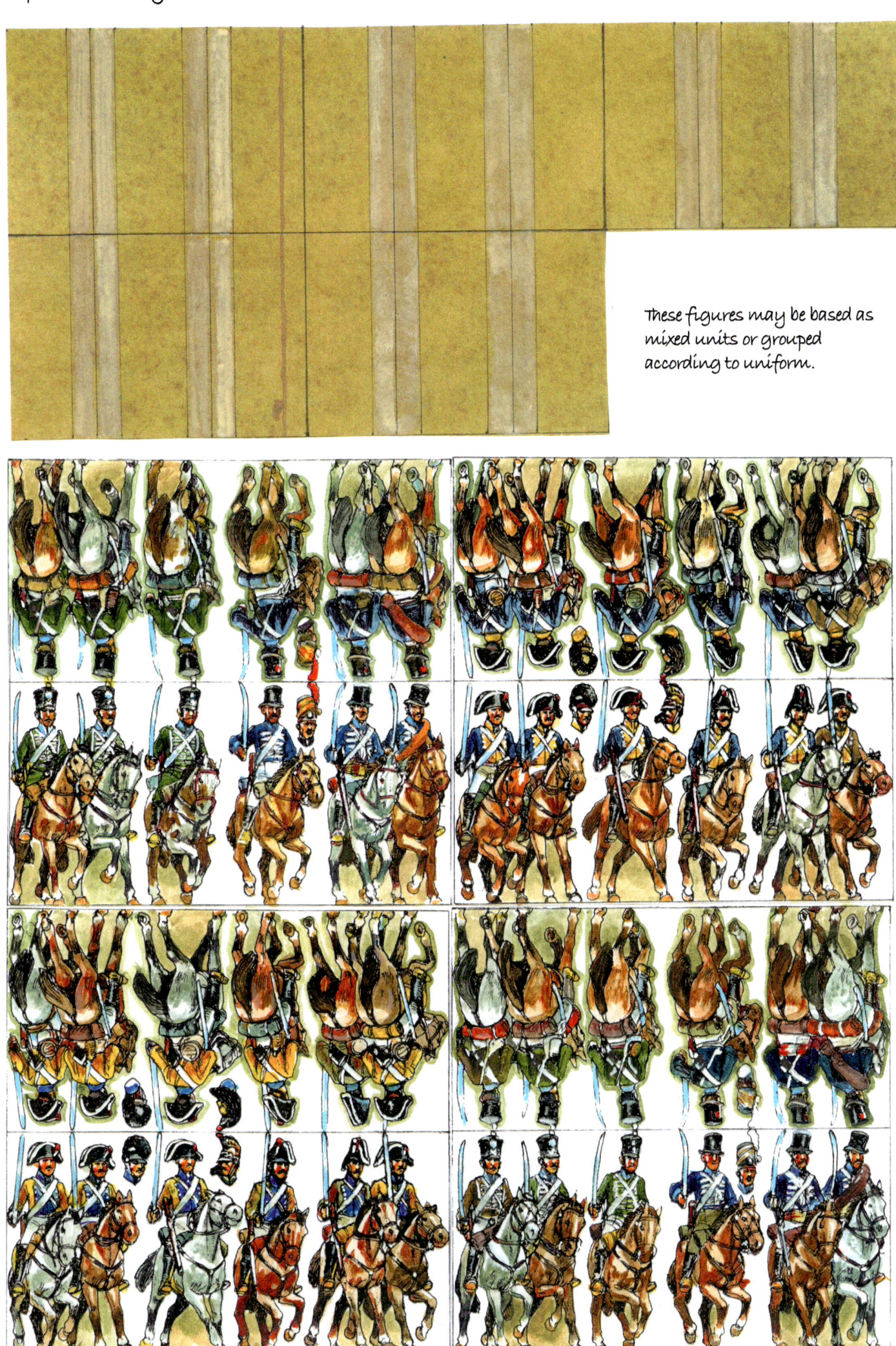

These figures may be based as mixed units or grouped according to uniform.

Spanish Guerillas (mount close order or as skirmishers on bases from page 47

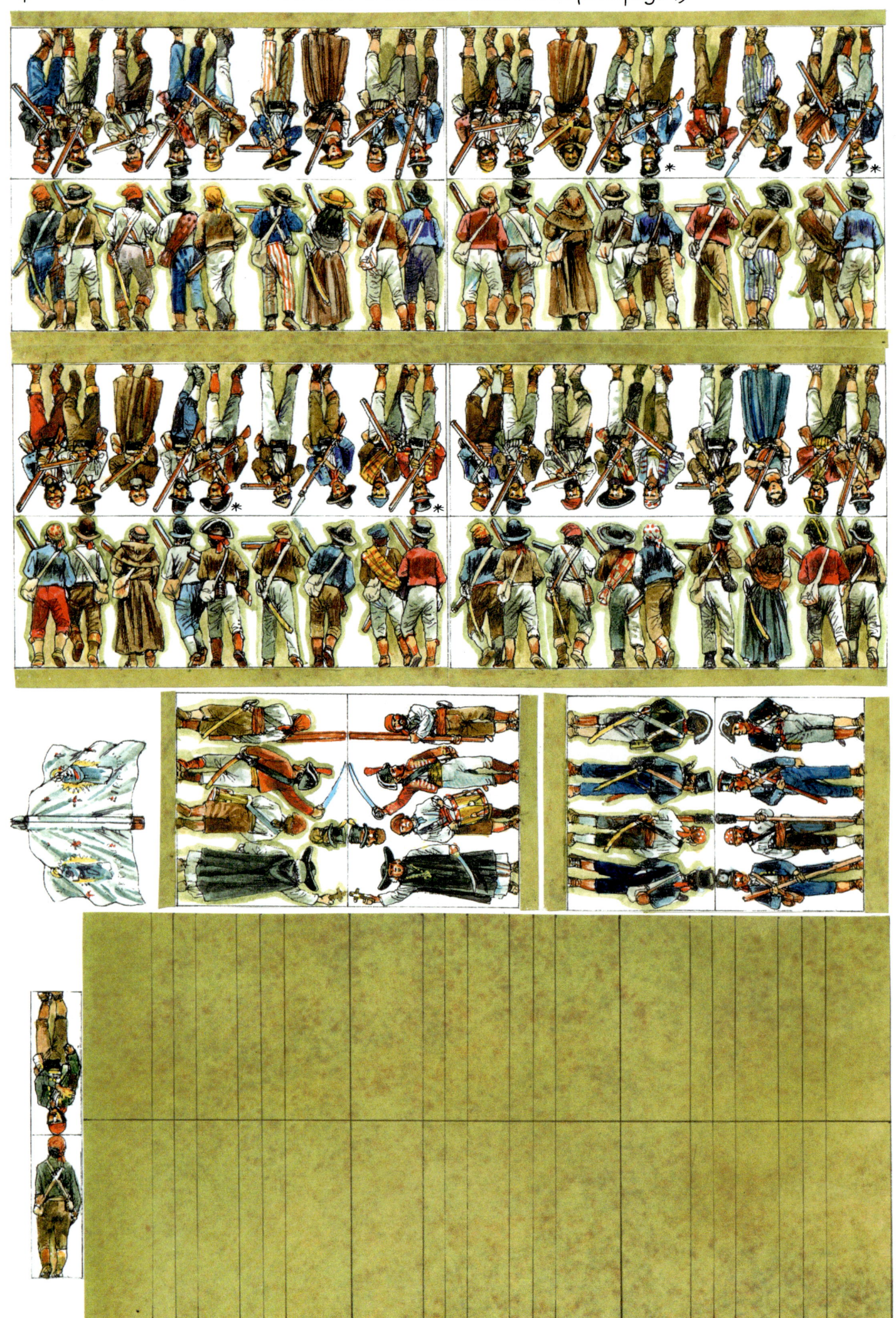

Spanish Guerilla Cavalry

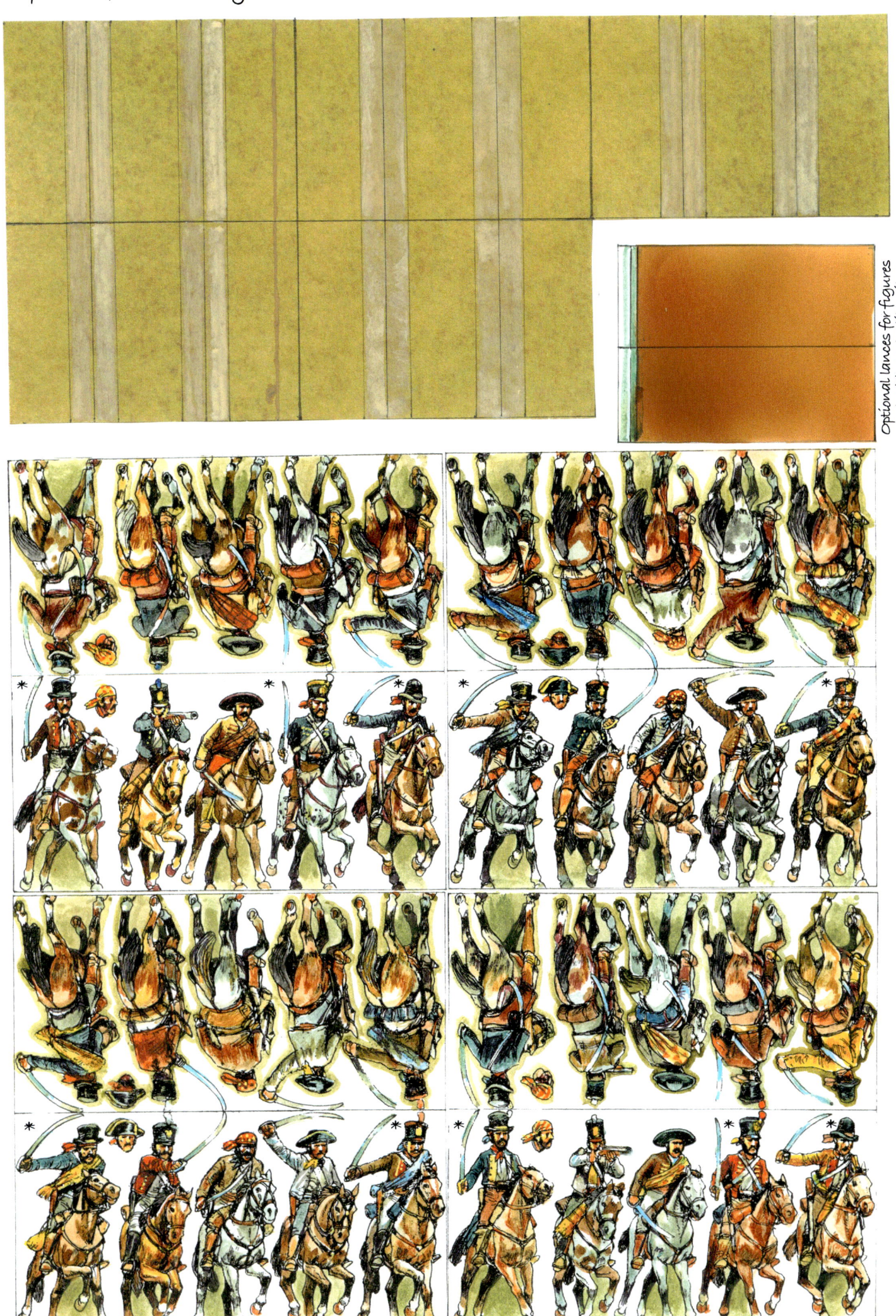

Optional lances for figures marked *. Make like pikes.

3D artillery, both sides

use bases from book.
See instructions from other sheets

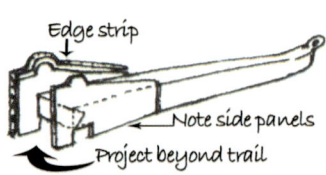

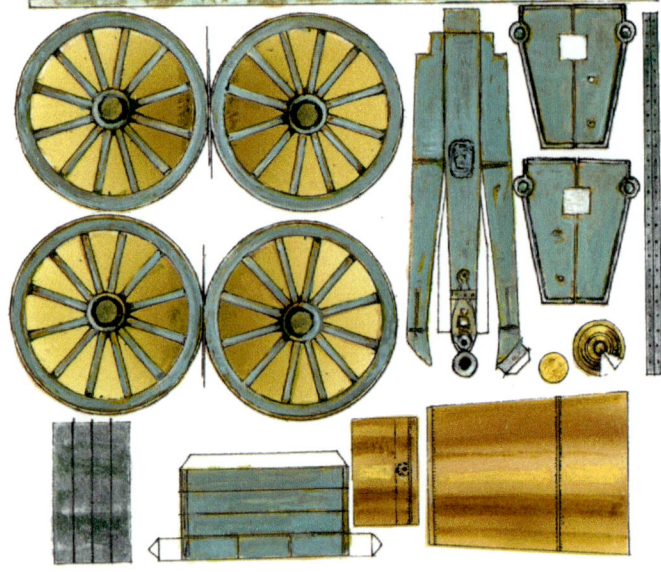

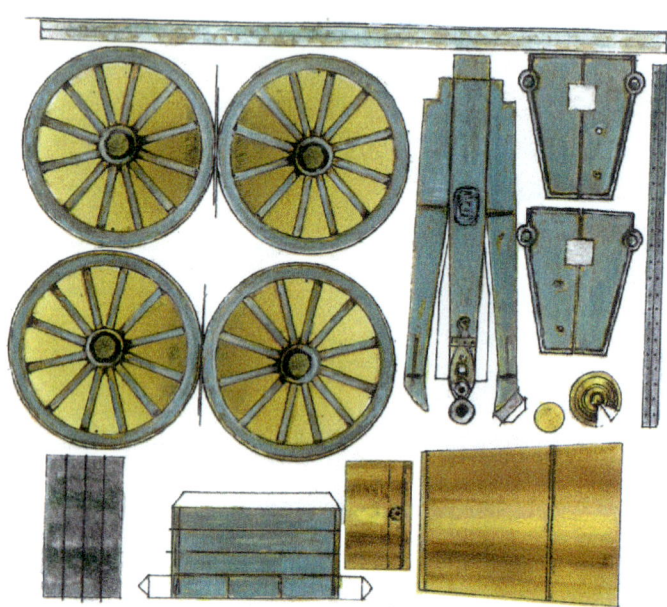

British and Portuguese guns

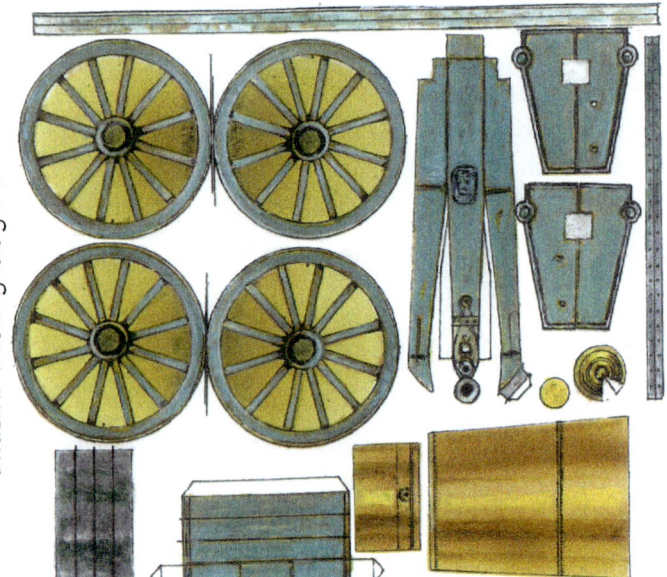

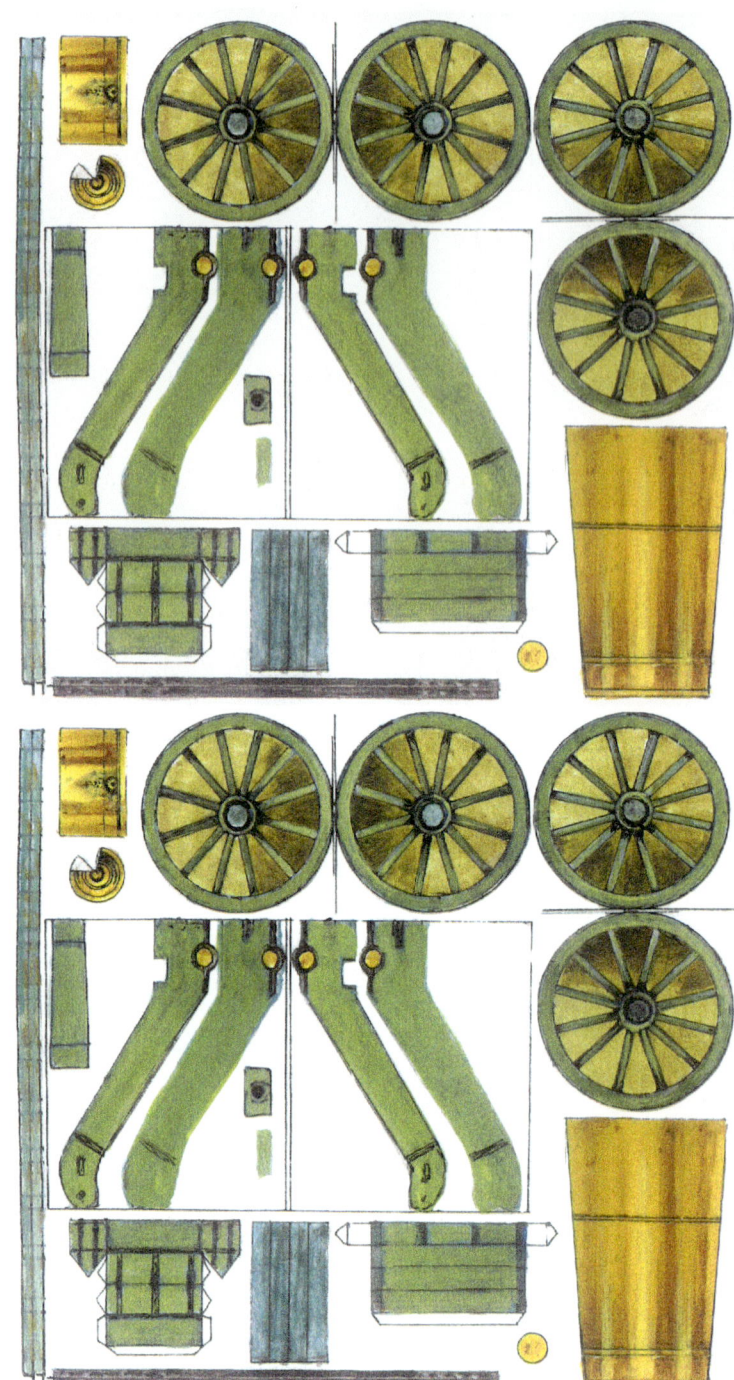

French and Spanish: cut barrel long for 12-pounder

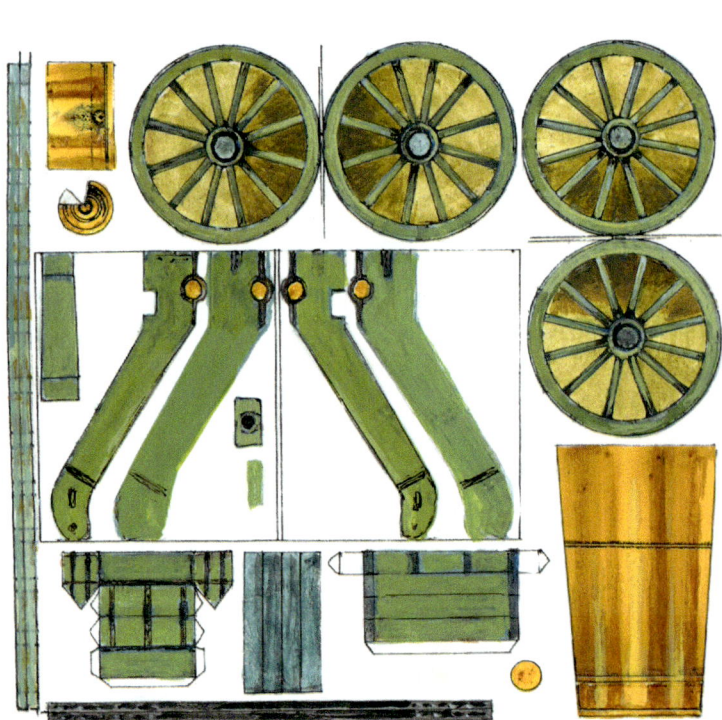

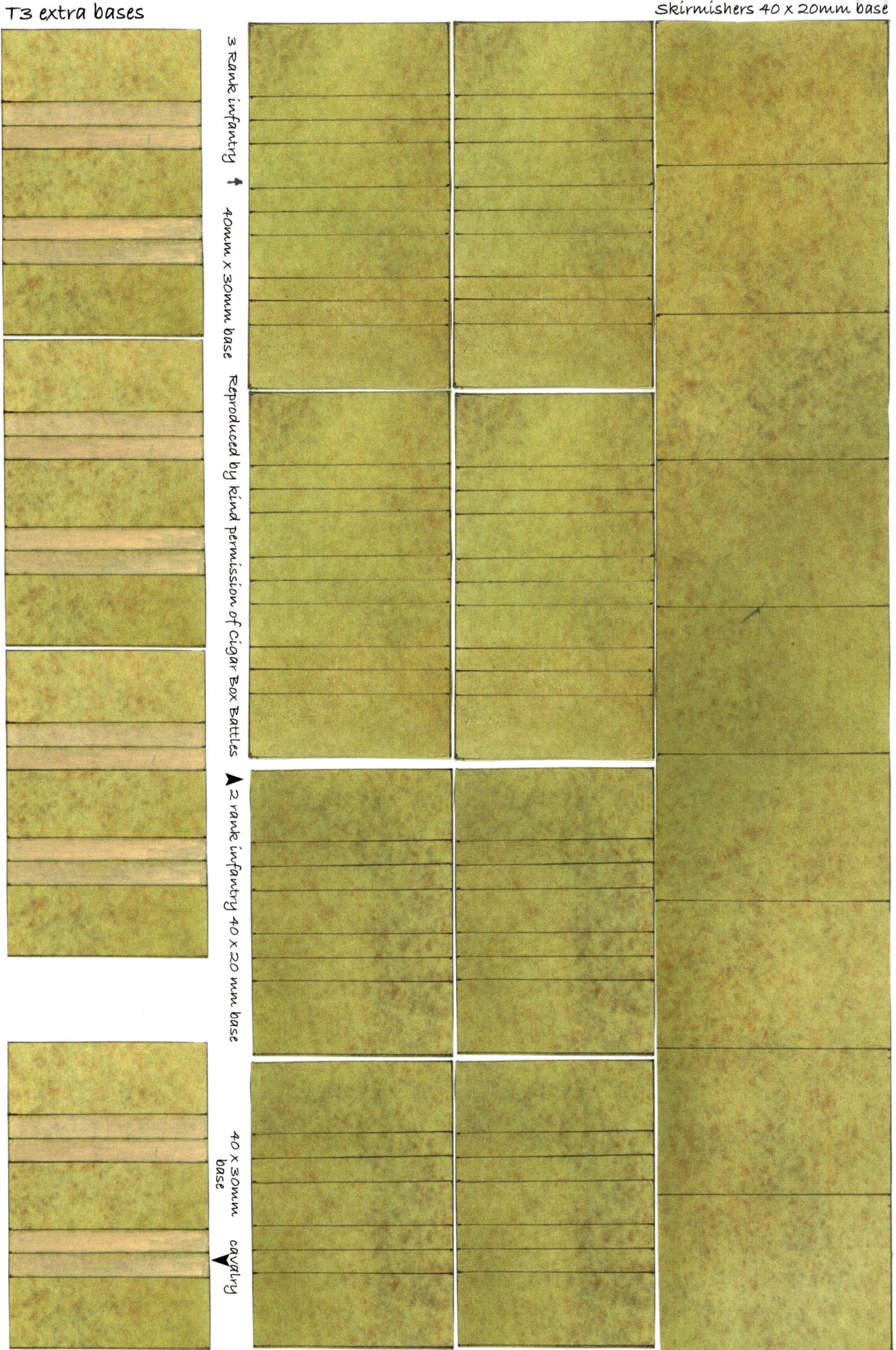

BATTALIONS AND FACINGS

A very basic guide to making armies from the soldiers in this book

Unlike our previous 'on Campaign' subject, The War of the Spanish Succession, the soldiers of the Napoleonic Wars are massively well described, with information about every uniform detail easily available. I have chosen to pitch the uniforms at around 1809. Many changes were made by all armies during the war, but often took many years to reach units out in Spain. This is a simple guide to get you started with the organisations and bits of hand-colouring you might want to do to make the regiments authentic. I like Staedtler fine-liner pens for this, but any very fine felt tip should do.

British infantry regiments were mostly single battalion. The collars and cuffs of their coats were called 'facings' and matched the field colour of the regimental flag except in the case of the Guards which have the crimson/purple flag. All regiments carried a 'King's colour' the union flag on the sheet, as well. The British centre companies (8 per battalion) are supplied with blue and yellow facings, the most popular colours. There is a sheet with white facings for white-faced regiments and to hand-colour for regiments with other coloured facings. The flank companies and skirmishers will need to have facings hand-coloured to match their parent unit. The light company and light infantry regiments were trained to fight in open order when required, as skirmishers.

All Guards, 'Royal' and King's German Legion regiments had blue facings. Here are a few other Peninsular regiments and their facings to get you going: Foot regiments: 3rd, 14th, 31st: buff (a colour like buckskin); 5th, 11th, 24th: green of various hues; 2nd foot: white.

Light infantry: 43rd: white; 51st, 68th: deep green; 52nd: buff; 85th: yellow. Highlanders: 42nd: blue; 79th: dark green; 92nd: yellow. Rifles: (specialist light troops armed with the long-range Baker rifle) 95th (2 battalions): black; 60th: red. These units mostly served in odd companies added to infantry brigades to boost their skirmishing power.

Cavalry was in short supply for the British early in the war. Most regiments were light Dragoons – the 11th is on the sheet with pale buff 'chamois' facings, the 14th had yellow facings and the 16th blue. Heavy Cavalry Dragoons: 1st and 3rd: blue; 4th: green. British cavalry at this time carried no flags on campaign.

Portuguese infantry were organised and mostly officered by the British in 2 battalion regiments. The 24 regiments followed a strict pattern in flags counting in threes. Ist: white field; 2nd: Red field; 3rd: yellow field; 4th: white – and so on. 1st to 12th had blue collars; 1st to 3rd: white cuffs; 4th to 6th: red; 7th to 9th: yellow; 10th to 12th: light blue. Cacadores: 1st to 3rd: brown collars; 1st: light blue cuffs; 2nd: red; 3rd: yellow. Cavalry 1st to 3rd: white collars and cuffs; 4th to 6th: red.

Spanish forces at this time were rather disorganised and wore uniforms from their old units or civilian clothing. The cavalry was in small units brought together where necessary to make regiments. The guerrillas were armed civilians fighting against the oppressive French regime and varied from highly professional soldiers to bandits.

French infantry and light infantry regiments fielded 2 battalions. The training and tactical use of line and light infantry was the same. For our purposes the uniform of all regiments was the same, so no hand-colouring is needed. They had allied foreign regiments with them in Spain some of which are in the book.

Polish Vistula Legion infantry. 2-battalion regiments. 1st: blue collar, yellow cuffs; 2nd: yellow collar and cuffs; 3rd: yellow collar, blue cuffs. They all had yellow lapels.

Swiss infantry. 2-battalion regiments with coloured lapels, collar and cuffs. 1st: yellow – you will need to use opaque model paint for this – 2nd: blue; 3rd: black; 4th: light blue.

French Cavalry. Note that French light cavalry did not carry flags in battle. Chasseurs' distinction system: regiments grouped in threes. The first in each group had distinction coloured collar and cuffs. The second had coloured cuffs but green collar and the third had green cuffs and coloured collar. Distinction colours: 1-3: scarlet; 4-6: yellow; 7-9: rose pink; 10-12: deep crimson; 13-15: orange. That's enough for our purposes. There were 31 regiments in all. Dragoons, mounted or on foot were numerous and active in Spain. They were grouped in sixes with some having green collars and cuffs and some in the distinction colour. All wore lapels in the distinction colour, which were: 1-6: scarlet; 7-12: crimson; 13-18: rose pink; 19-24: yellow; 25-30: orange. You can choose to have units with epaulettes or helmet covers. Campaign dress varied a lot.

Please note that further troops for both sides will be available on www.peterspaperboys.com and you can keep up with developments in all periods of Paperboys on Facebook: the Paperboys Page. A brief expansion set to update The War of the Spanish Succession rules will be available for free download from the Helion & company website Paper Soldier section.